AN INTRODUCTION TO THE
SCIENTIFIC STUDY OF THE SOIL

An Introduction to the Scientific Study of the Soil

Fifth Edition by
W. N. TOWNSEND, B.Sc.
Senior Lecturer in Agricultural Chemistry in the University of Leeds

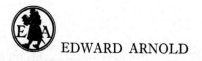

EDWARD ARNOLD

© W. N. Townsend, 1973

First published 1927
by Edward Arnold (Publishers) Ltd
25 Hill Street,
London W1X 8LL

Reprinted 1929
Second edition 1932
Third edition 1936
Reprinted 1940, 1943, 1945, 1946, 1948
Fourth edition 1960
Reprinted 1964
Reprinted 1968

Fifth edition 1973

Boards Edition ISBN: 0 7131 2407 5
Paper Edition ISBN: 0 7131 2408 3

Made and printed in Britain by
Butler & Tanner Ltd, Frome and London

Preface

In this edition the aim of the original author has been preserved, and the book is 'essentially a communication from a teacher to students'. It is hoped that students interested in scientific aspects of agriculture will be given an insight into some of the factors which have an important bearing on soil fertility and will be stimulated to delve more deeply into the detail of those areas which they find of special interest.

A number of themes have been given a revised presentation with a view to updating current ideas although the basic pattern of the book has been retained.

It is impossible to acknowledge individually all who have helped, directly or indirectly, with this volume, it is hoped they will identify themselves and accept the author's grateful thanks.

Leeds, 1972   W. N. T.

Contents

I   The soil—plant complex

The study of the soil may be pursued for a variety of reasons. Soil is a natural material and merits attention in its own right. It is also the material supporting the growth of higher plants which are essential for the health and well being of man, providing the raw materials for much of his food and his clothing. In this connection knowledge of those soil properties affecting plant growth is supremely important and it is hoped that the following chapters will contribute to a more general appreciation of the soil and its influence on vegetation as a whole.

Plants and Plant Nutrition

One distinguishing feature of a plant as opposed to an animal is its immobility. This means that its nutritional requirements to ensure its development and survival must be available where it grows. These essentials are normally obtained from the soil. However, development and survival of individual plants do not in themselves constitute sufficiency. Maintenance and evolution of species require that individuals reproduce, and the full life cycle involves a sequence in which vegetative growth, flowering development, seed formation, seed germination and growth, give a succession of plant generations. Some plants can reproduce by vegetative propagation. This involves the development by some component part of the adult plant of a new individual possessing the full parental characteristics and having the potential for independent existence, if and when it is separated from the parent.

Manifestations of the various methods of plant reproduction exhibit many differences in detail, but the fundamentals of a true life cycle for all plants irrespective of species involve:

(a) **Germination,** when the dormant seed is stimulated to develop and the growth process starts. During this phase the growth requirements are met by internal reserves of food, the only external requirement being water. If the environment is congenial the seed develops its plumule, which is the precursor of the aerial vegetative parts of the plant, and radicle, which is the precursor of the root system. The fact that the seed itself is meeting its growth requirements both in terms of raw materials and energy is confirmed by the fact that the substance of the plant, as measured by its dry matter content, actually decreases as germination proceeds, water absorption alone accounting for the increase in the gross weight of the seedling.

The development of the plumule above ground allows initiation of the process of photosynthesis, whereby light energy is harnessed by chlorophyll to promote growth processes, and the appearance on the radicle of true roots enables constituents of the soil solution to be absorbed, all of which lead to the next stage of growth:

(b) **Vegetative development.** This phase lasts for a variable time, in some plants culminating when the seeds are produced or in others running continuously and supporting the seasonal production of fruiting bodies over many years. During vegetative development carbon dioxide is absorbed from the atmosphere by the aerial tissues and is elaborated into carbohydrates, using light energy to motivate the reaction. These organic compounds react within the plant with other elements or compounds which have been absorbed or elaborated *in situ* and the complex structures which constitute the whole plant are formed. Energy to motivate these further reactions is obtained by oxidation of some of the carbohydrates resulting from photosynthesis by the process of respiration. Respiration is a continuous process requiring an oxygen supply and evolving gaseous CO_2 as an end product. Photosynthesis elaborates CO_2 into carbohydrate, releasing O_2 as a by-product, but the process can only take place under the influence of light of appropriate wavelength. In plant growth the net intake of CO_2 must be positive as this is the principal source of organic carbon for plant tissue development. The photosynthetic reaction must therefore have the dominant role, although during hours of darkness loss of CO_2 due to respiration will take place.

The movement of metabolites (the products of metabolism) within the plant, takes place through the various conducting tissues with water acting as the vehicle for transport and also taking part in the reactions concerned. Within the growing plant there is a continuous flow of water, the overall movement being from root to leaves. One of the most important functions of the plant root is the absorption of water, some of which eventually evaporates from the leaf surfaces, the whole process being referred to as transpiration. The continuous maintenance of transpiration is essential if plant growth is to be maintained without check until the final phase of growth is achieved, i.e.:

(c) **Maturity,** when vegetative development ceases and seed is set, or food materials are conserved in special storage organs for future vegetative development.

Near maturity the pattern of plant metabolism changes. Photosynthesis declines and respiration becomes the dominant process, dry weight changes become zero or negative, and translocation of metabolised products takes place within the plant without any further call for nutrients. The leafy tissues become senescent and are discarded by the plant which may die if it has fulfilled its life cycle completely or, as is the case for many plants, enter a resting period. Renewed activity in the next growing season will be initiated by demands upon the food reserves laid down during earlier growth.

The processes described above take place as a result of a series of complex interrelated chemical reactions which occur within the plant cells and which are activated by enzymes. These are essentially reaction catalysts, although their action is more than simply catalytic as they can both initiate a reaction and exercise control over its rate. They may be defined as the functional units of chemical activity and are essential to all living processes. There are many enzymes involved in the development of living organisms, each having a unique and specific role to play with particular reference to its own reaction component. They are given names descriptive of the type of reaction they control, all the names ending with the suffix -ase. One group of enzymes, for example, effects hydrolyses, and its constituent enzymes can be classified according to the sort of compounds they attack, e.g. proteases hydrolyse proteins, peptidases break down peptides, carbohydrases decompose carbohydrates and include the amylases which break down starch; cellulase, hemi-cellulase, invertase, cellobiase are further examples. Peroxidases, hydrogenases and dehydrogenases amongst others all perform the reaction functions which are reflected in their names.

The structures of these enzymes and the way they work is not perfectly clear, but a great deal is known about them. Fundamentally they are proteins but before they can become active, association with some prosthetic group is required. A specific heavy metal or phosphoric acid is often contained within the prosthetic group and has an important bearing on the enzyme activity.

It is obvious that if a plant is to grow successfully all the life process mechanisms must work. This means that all the necessary materials for its nutrition must be available to it, its environment must be suitable, and no toxic materials should be present. Satisfactory growth is not easy to define; it does not necessarily mean maximum growth. Plants growing under natural conditions may often be stimulated to greater development by applying fertiliser, or by giving more water, or as a result of warmer weather. Plants grown for food or other human necessity are normally grown under conditions of some economic control and can often be stimulated by suitable treatment to give above average yields, although the law of diminishing returns may soon operate and the cost of stimulating yield may become greater than the return from the extra growth obtained. As the growth factor that can most easily be varied is a nutritional one a great deal of research has been carried out on the nutritional requirements of crop plants, and some of the conclusions reached are discussed later. Before any considerations of quantitative nutrient requirements can be made a qualitative survey must be carried out in order to determine what the requirements actually are.

Plant Analysis

A first and obvious approach to the problem of defining the nutritional requirements of plants is to subject them to chemical analysis. Indeed the first rational advice on the use of agricultural fertilisers was made on the basis of plant analyses in the early years of the 19th century and controversial issues were opened between Justus von Liebig, the German chemist, and Mr Lawes and Dr Gilbert in England on the interpretation of the chemical analyses of plants and fertiliser practice.

It is possible to regard plants as mixtures of chemical compounds, and to analyse them as such, but unfortunately there is no finality in plant composition. Variations occur between different crops growing in similar environments and similar soils, between the same crop growing in different environments and in different soils and between different plants of the same crop of different ages. No quantitative interpretation of chemical analysis can, therefore, be firmly established. Examination of many hundreds of results can, however, enable several qualitative conclusions to be drawn about plant composition and the many experiments that have been made relating changes in nutrition to yield and other characteristics have led to much valuable knowledge.

While approximately 90 per cent of the dry matter composition of plants is made up from the elements carbon, hydrogen and oxygen, and perhaps nitrogen, which are traditionally associated with organic compounds, the remainder consists of inorganic elements. These are introduced naturally into the plant from the soil in which it grows and are wholly absorbed by the root systems in the first instance (see Table 1). In this respect they are similar to nitrogen which is

TABLE 1
THE APPROXIMATE COMPOSITION OF SOME COMMON CROPS
(all figures are given in round numbers)

CROP YIELD (Kg/ha)	WHEAT Grain 3,750	WHEAT Straw 2,000	BARLEY Grain 3,500	BARLEY Straw 1,700	TURNIPS (Roots) 42,000	MANGOLDS (Roots) 62,000	POTATOES (Tubers) 24,000	MEADOW HAY 3,200
Approximate Composition (%)								
Water	13	14	15	14	90	88	76	14
Oil	2	11	2	2	1	—	—	3
Crude Protein	12	2	9	3	1	1	2	9
Sol. Carbohydrate	69	40	68	42	6	9	20	41
Crude Fibre	2	36	5	34	1	1	1	26
Ash	2	7	1	5	1	1	1	7
Ash Composition (%)								
P	22	1	15	1	5	5	7	3
K	27	13	26	17	42	42	50	20
Ca	2	3	2	7	9	3	2	10
Mg	6	1	5	1	2	2	3	4
Na	1	—	1	4	6	14	2	4
Fe	1	1	—	1	—	—	1	1
S	1	2	1	3	6	2	3	4
Cl	—	2	2	7	5	4	3	7
Si	—	32	9	20	—	1	1	14
Removals from soil by whole crop (Kg/ha)								
P	17		10		12	24	12	6
K	40		32		115	210	75	47
Ca	12		8		23	19	2	23
Mg	9		5		6	10	5	10
Na	3		5		17	83	2	8
Fe	3		2		2	2	2	2
S	19		10		17	9	4	9
Cl	5		6		14	22	4	17
Si	9		38		2	4	2	33
N	66		55		55	93	50	55
Transpired Water	150,000		140,000		100,000	225,000	112,000	90,000
CO_2 utilisation	750		700		700	1,250	500	500

also introduced through the root system, but they are dissimilar to carbon, hydrogen and oxygen which, although they can be introduced through roots via bicarbonate ions and water molecules, can also be obtained from atmospheric sources such as carbon dioxide and water vapour.

The inorganic content of plants is often collectively measured and recorded as 'ash' which is the residue after ignition in air, when the organic compounds are completely decomposed and volatilised. Analysis of plant ash using modern analytical techniques shows that practically every element in the periodic table can be present although the amounts of some elements may be infinitesimally small. This does not in itself mean that all these elements are necessary for the healthy growth and development of any one plant in particular or plant life in general. Some of the elements may only be present in plants growing in particular environments and may be absent from plants of the same species growing elsewhere, but there are many elements which are invariably present in all plants of all species and this suggests that these may play some essential part in growth processes. However, the only way to confirm whether or not such elements are truly essential is to grow plants in media from which they are individually excluded. If any element is necessary for growth, its exclusion will drastically impair plant development or even cause growth to cease altogether. For several elements proof of essentiality in plant growth processes generally has been firmly established; there is suggestion that certain plant species have particular additional requirements. There is still uncertainty about some elements because of the extreme experimental difficulties in excluding them absolutely from a growing medium, when the seed or original plant cutting may contain enough of the element for adequate growth, or difficulties in purifying the essential elements to a sufficient degree of refinement.

The inorganic elements known to be generally essential for plant development are K, P, S, Ca, Mg, Fe, Mn, B, Mo, Zn, Cu and Se. While the absence of any one of these elements will prevent proper plant growth, the quantitative requirements of each one differ quite considerably and this enables the list to be divided into two main groups.

In the first group are those elements required in significantly large quantities and which, within certain limits, show a direct correlation between amount of growth and quantity of nutrient element available. These elements usually occur as constituent elements in organic compounds elaborated during growth, e.g. Magnesium is a constituent of chlorophyll; Phosphorus and Sulphur are constituent elements of certain proteins but they may also be concerned with the actual syntheses of such compounds during growth, e.g. Phosphorus is involved in high energy chemical bonds which enable certain reactions to proceed in a forward direction; Calcium can combine with surplus organic acids which might slow down metabolic processes and release them should there be a demand for any particular supply. The elements in this first group are K, P, S, Mg, Fe and Ca, and these are often referred to as macro-nutrients.

In the second group are the elements which are only required in extremely small amounts for full and successful plant growth, and provided a minimal supply is available to the plant, growth proceeds to the maximum extent. These elements have been shown to be the activators of the enzymes as mentioned earlier and in many enzyme systems specific relationships exist although in some enzymes a certain degree of substitution of metal-coenzyme can be tolerated. These trace elements or micro-nutrients include Mn, B, Mo, Zn, Cu, and Se. As well as the micro-nutrients essential for plants, certain additional elements are required by animals and, as grazing animals normally exist on plant products, their presence in the food is necessary. In this category are cobalt, iodine and fluorine, and in the agricultural production of crops it is important that these elements should be present in the plant in adequate amounts.

Some elements inhibit biological growth, and their absorption by plants can cause death or lack of perfect development; arsenic, chromium, lead, and nickel are in this category, but it is also important to note that several of the essential nutrient elements can also be toxic to the plant or to the grazing animal if they are absorbed by the plant in amounts in excess of normal requirements. Of particular importance here are boron, copper, fluorine, molybdenum and selenium. All the other elements which may or may not be present in plant ash are there accidentally. Their adventitious presence in the soils leads to absorption by the plant root and their inertness to biological reaction invokes no special mechanism on the part of the plant to reject them.

The following list gives the status of various elements in plant nutrition, including those elements essential for animal growth, which normally originate in plants. Elements of doubtful importance or of importance in a limited range of plants are placed in brackets.

Macro-nutrient elements		Micro-nutrient elements		Harmful elements	Incidental elements			
C	Mg	B	(Ba)	As	Ag	Ge	Pl	Ti
H	S	Co	(Ga)	Cr	Au	Hf	Pt	Tl
O	Ca	Cu	(Li)	Ni	Be	Hg	Re	W
N	Fe	F	(Ra)	Pb	Bi	In	Ru	Y
P	(Si)	I	(Rb)		Cd	Ir	Sb	Yb
K	(Na)	Mn	(Sc)		Ce	La	Sm	Zr
		Mo	(Sr)		Cs	Nb	Sn	
		Se	(U)		Dy	Nd	Ta	
		Zn	(V)		Eu	Os	Te	
					Ga	Pd	Th	

Soil Studies and Plant Growth

Soil is not just a simple mixture of inorganic minerals with organic matter in which plants grow. From it plants obtain their mineral nutrient needs and the water necessary for growth and transpiration. The soil/root relationship is an intimate one and the ways in which materials are transferred from the one medium to the other are complicated. Physical, chemical, and biological systems are involved and some understanding of these sciences is necessary to appreciate the problem of soil fertility. The complex role of the soil in agriculture has long been appreciated.

It was at one time supposed that the scientific study of the soil was mainly a matter for chemists, and that the chemistry involved was fairly simple and straightforward. The following chapters will show that the chemistry of soils is far from being simple and straightforward, and that in spite of the big advances made during the last twenty-five years or so, a great deal of the fundamental chemistry of the soil is still obscure. Indeed it is quite likely that the state of our knowledge of general chemistry is a limiting factor to the development of soil chemistry. Almost every new development in physical chemistry seems to throw fresh light upon soil problems.

As men of science gave more attention to the soil, not only did it become apparent that the higher ramifications of physical, colloid and organic chemistry are involved, but also that microbiology and physics are involved along with chemistry in the study of soil fertility.

Pedology

It was not unnatural, therefore, that the segregation of the science of the soil as a special subject, with a name of its own, was suggested in various quarters. The suggestions first put forward did not receive a very warm welcome and were, in fact, opposed by some students of the soil in this country. Such opposition was quite sound, for the application of a particular science or several sciences to an industrial problem does not necessarily constitute a new subject. Nothing can really attain the dignity of a 'subject' as the word is understood in scholastic curricula, until it has something fundamental which it can call its own. The mere application of chemistry and physics to minerals and rocks does not make geology, and the application of those sciences to the soil does not make a new subject.

In recent years, however, the outlook has been considerably changed by the outstanding work of the Russian school. It may be fairly claimed that the study of soil formation processes and soil types, which is briefly discussed in later chapters, has given to Soil Science a fundamental characteristic which had been properly recognised by the adoption of the word Pedology,[1] to describe the science of the soil. The work of Sibirtzev, Glinka, Stebutt and others put a backbone into the study of soils and turned that study into Pedology, in the same way that William Smith put the backbone into the study of rocks and made it Geology.

The word Pedology has been almost unanimously accepted on the Continent and to a large extent in America. Its use is now quite common in this country although there is a tendency to confine it to the study of soils in the field, and to consider the laboratory studies of soils as being soil chemistry or soil physics or just soil science.

[1] From the Greek πεδου—soil.

2   The mineralogy of soil

With the exception of a few organic soils, the bulk of soil material is mineral in character, and has been derived from solid geological deposits. Rock formations must however, undergo considerable modification before their material can become a corporate part of soil. While it is true that the mineral constituents of some mature soils do not necessarily bear a close similarity to those of the rocks from which they originally came, parent rocks do impress upon many soils features which deserve attention.

The rocks of the earth's crust fall into two main categories: igneous rocks which are formed by solidification of magma, and which are, with the exception of a few vitreous types, solid aggregations of a range of mineral species; and sedimentary or fragmental rocks which have been derived from igneous, or from preformed sedimentary rocks, largely through the agency of water. These latter occur in the main as consolidated rocks although they may take the form of unconsolidated deposits. These are also composed of mixtures of mineral species but differ quantitatively from the parent formations due to chemical and physical changes induced by aqueous and atmospheric agencies.

Igneous and sedimentary rocks may suffer *in situ* alteration by heat and/or pressure consequent upon geological activities. These are metamorphic rocks and their constituent minerals usually undergo appreciable change during the metamorphosis.

Minerals Occurring in Rocks and Soils

The solid state which characterises the accessible parts of the earth's solid crust is essentially a crystalline state although there are solid components which fail to reveal any ordered molecular structure and therefore must be classed as amorphous. This solid crust is heterogeneous and made up of units which in themselves possess an ordered atomic arrangement which has a constant and repeating pattern throughout the unit. These are the crystalline minerals, which number some thousands. However, only a few dozen species of crystalline minerals figure to any significant extent in the composition of the earth's crust.

It is also noteworthy that significant crystalline minerals constituting the bulk of the earth's crust are made up of very few chemical elements. It has been computed that of every 100 atoms in the solid crust, of the order of

$$60 \text{ will be oxygen}$$

20–21	,, ,,	silicon
6–6·5	,, ,,	aluminium
2·5–2·8	,, ,,	sodium
2·5–2·8	,, ,,	hydrogen
1·4–1·9	,, ,,	calcium
1·4–1·9	,, ,,	iron
1·4–1·9	,, ,,	magnesium
1·4–1·9	,, ,,	potassium

No other element approaches 1 atom per cent on average although of course local concentrations occur to modify such a generalisation.

This numerical distribution means that any orderly or crystalline arrangement is dominated by oxygen atoms and that in any systematic geometrical arrangement all the other elements will be surrounded by oxygen atoms. It is also possible to conceive of the various rock-forming minerals as being built up of mixtures of oxides e.g. ortho-clase felspar $2KAlSi_3O_8 = K_2O.Al_2O_36SiO_2$ and, especially in the older literature, mineral formulae are presented in this form.

The fact that silicon occupies second place in the quantitative scheme of elements suggests that silicon-oxygen combinations, which will be anionic in character, will be prevalent and that these with the oxides of the metals, which will be basic, will form neutral or salt-like compounds, and that associations of this kind will determine the structural characteristics of the minerals concerned. This is in accord with an assessment by Clarke that of the minerals present in the earth's crust, silicates predominate. The relative proportions of the principal types are:

$$59\cdot5\% \text{ felspar varieties}$$
$$12\% \text{ Quartz}$$
$$16\cdot8\% \text{ Pyroxines and Amphiboles}$$
$$3\cdot8\% \text{ Micas}$$

SILICATE MINERALS

The Felspar Minerals

The felspars comprise a group of alumino silicates combined variously with potassium, sodium and calcium, the idealised members being $KAlSi_3O_8$, which may be either *orthoclase* or *microcline* according to whether the crystal system is monoclinic or triclinic, $NaAlSi_3O_8$ *albite* and $CaAl_2Si_2O_8$ *anorthite*. Albite and anorthite are mutually soluble and form a continuous series of minerals known as the plagioclase felspars.

The importance of the felspar minerals in soil studies lies in the fact that they undergo changes when exposed to weathering influences, and give rise to a number of alteration products, many of which ultimately form important constituents of soils. Conditions bringing about the decomposition of the felspar minerals are complex and not easily definable, although hydrolysis reactions dominate the breakdown. Amongst the products formed in different circumstances may be mentioned; the clay minerals, especially kaolinite, pyrophyllite and halloysite; secondary minerals such as the zeolites, muscovite mica and glauconite; and simple minerals such as quartz, calcite and gibbsite. Decomposition is also accompanied by the liberation of alkali cations.

Amphibole and Pyroxene Minerals

The pyroxene group of minerals and the amphibole minerals, of which *hornblende* is a characteristic member, differ from the felspars in that they are ferro-magnesium silicates. A number of individuals contain also calcium and more rarely sodium. Both these groups of minerals decompose comparatively easily under normal weathering conditions, the products of decomposition following similar trends and including biotite mica, quartz, calcite, and particularly chlorite minerals which are hydrous silicates of magnesium and ferrous iron, and which frequently contain aluminium. Oxidation of the iron easily occurs in the chlorite crystals.

Micas

These minerals occur widely in igneous rocks and in many sandstones and shales. There are two principal groups, *muscovite* and *biotite* micas. The former are essentially colourless minerals which are represented by an idealised formula $Al_3KH_2Si_3O_{12}$, whilst the latter are dark coloured with an idealised formula $Al_2Mg_2KHSi_3O_{12}$. A number of variants of these characteristic forms are recognised which contain modifications in the cations involved in their structures. They decompose relatively easily, biotite more so than muscovite.

Quartz

Quartz, or crystalline SiO_2, is very resistant to chemical breakdown, and upon the disintegration and weathering of the rocks in which it occurs suffers only physical change and it is therefore almost universally present in soils. It may constitute up to 99 per cent of some sedimentary rocks.

NON-SILICATE MINERALS IN ROCKS AND SOILS

In addition to the various silicates many other minerals are found and, although present in relatively small amounts, these have important influence on soils in terms of both forms and properties. Oxides, hydroxides and carbonates are the main chemical types but others such as phosphates and sulphates are often present.

Iron Compounds

There are six naturally occurring iron oxides or hydroxides found in rocks and soils. Some are primary minerals, existing in the original rock, others are secondary being the result of weathering processes. They show differences in colour and because of this can often be used to interpret patterns of soil behaviour.

Magnetite Fe_3O_4 is black, and *haematite* Fe_2O_3 is red. These are usually regarded as primary minerals and are found as discrete crystals in igneous rocks. *Maghemite* Fe_2O_3 is a dark brown material usually found as a secondary mineral and often in association with organic matter. While it is crystalline it is not usually found as well developed crystals but rather as a diffuse component of the upper horizons of the soil.

Two other crystalline iron minerals *geothite* and *lepidocrocite* are found as coating or concretionary components. These are hydrated iron oxides with the empirical formula FeO.OH. They crystallise slightly differently, *geothite* with a hexagonal and *lepidocrocite* with a cubic lattice. The distinction is important as the cubic lattice indicates that iron in the ferrous state must have been involved at a precursive stage of development, a state often associated in soils with waterlogging. Both minerals are yellow-brown in colour, goethite being somewhat yellow and lepidocrocite somewhat orange.

A more generally widespread hydrated iron hydroxide
$$(Fe(OH)_3nH_2O)$$
occurs also as a secondary coating material of many soil particles and

is also found adsorbed on clay. It is amorphous and yellow in colour and is referred to as hydrated ferric oxide gel or *limonite*.

Other Minerals

Aluminium hydroxide $Al(OH)_3$ occurs in rocks and soils as *gibbsite*, but it is also found, often associated with hydrated ferric oxide gel, as an amorphous hydrated material diffusely adsorbed onto the surfaces of clay mineral particles. In this case it is always secondary and in some pedological situations can, again like its iron analogue, be transported within the soil profile to give certain characteristic soils.

Oxides of manganese are often to be found in rocks and soils, usually in situations where iron compounds are prominent. The most commonly occurring form is MnO_2, *pyrolusite* but in soils identification is often very difficult because of the existence of many derivatives formed by hydration and the substitution of Mn^{4+} by Mn^{2+} and other cations.

Of much greater significance are the carbonates of calcium and magnesium. Many secondary rocks are predominantly calcium carbonate in the form of calcite ($CaCO_3$) or calcium magnesium carbonate ($CaCO_3MgCO_3$) *dolomite*. These are the limestones, calcareous or dolomitic; or the chalks and many give rise directly to soils.

Rock phosphates occur widely as *apatite* minerals. These have a general formula $3Ca_3(PO_4)_2.CaX$, where X may be CO_3 (carbon-apatite), $(OH)_2$ (hydroxy-apatite), F_2 (fluorapatite), Cl_2 (chlor-apatite). They are important as being the major source of mineral phosphorus. The sedimentary rocks in which they occur in massive form are of marine origin. The phosphorus occurring in soils is so diffusely distributed that its mineralogical composition is somewhat speculative but it does occur associated particularly with calcium, iron and aluminium.

Occurring as definite and individual minerals in both rocks and soils is a number of so called accessory minerals. These are very hard and resistant species, not susceptible to chemical change. They are often used as marker or reference minerals to enable quantitative measurements of rates of change in other associated materials to be assessed. They include magnetite and hematite, which have been mentioned earlier, together with *rutile* TiO_2, *ilmenite* ($FeTi_2O_3$), *zircon* ($ZrSiO_4$), *garnet* ($CaFe_2Si_3O_{12}$) and *tourmaline*, a complex borosilicate which is also the primary source of the essential plant trace element nutrient, boron.

Rock Weathering

The transition of rocks and their constituent minerals to soil is not simple, and involves a complex series of changes, usually referred to as weathering. This involves physical and chemical alterations and gives a mineral complex of soil parent material upon which a soil develops following the incorporation of organic matter and the development of a biological population of a diverse kind. The latter includes bacteria, fungi, algae, protozoa and several genera of the invertebrates and even the vertebrates.

The weathering changes involved in the formation of soil parent material can be considered under two broad headings, namely physical weathering and chemical weathering; but while these are discussed separately for convenience, they occur simultaneously in nature and are to some extent interdependent in that the rate of chemical change is influenced by the degree of physical comminution. The ease of disintegration may be affected also by the extent of chemical change that has taken place.

Physical breakdown is simply a splitting down of rock masses with the production of rock fragments and implies no change of composition. It is brought about by one or more of a number of natural agencies including the action of wind, water, ice, temperature change and growing plants. Water and wind exercise an abrasive action on rock masses by virtue of solid particles carried by them which wear down the surfaces upon which they impinge. Ice masses in the form of moving glaciers also grind down the surfaces over which they pass, the actual grinding agent being not so much the ice itself but the rock debris picked up as it moves over the ground. The conversion of water from liquid to solid ice at 0°C. is accompanied by a volume expansion of approximately 9 per cent, and this expansion is responsible for much disintegration of rock masses (frost weathering), by reason of the pressures developing when the water in hair cracks and crevices in the rocks is brought below freezing point. Rapid temperature fluctuations may also be responsible for a certain amount of disruption (insolation weathering) of rocks composed of a heterogeneous mixture of different mineral species each having a different thermal coefficient of expansion. Repeated heating and cooling of such rocks leads to stresses and strains within the structures, which in time develop planes of weakness leading to an exfoliation of the surface layers. Plants add to the rock-breaking factors through their root development, and many instances have been recorded of tree roots in particular, penetrating small cracks in rocks, growing and eventually splitting off large boulders.

Chemical weathering processes are much more important than simple physical disintegration in so far as soil studies are concerned, as the soil embodies the resultant products. Several agencies are again involved and in any one instance the different factors act together, giving a chain of chemical reactions that is difficult to follow precisely. In the main, the bulk of chemical weathering is brought about by water, with or without the presence of carbon dioxide, and by oxidation.

The action of water is threefold. It may bring about hydrolysis, hydration or simple solution. Hydrolysis may be accomplished by water alone, but as is commonly the case, the reaction is catalysed by the presence of hydrogen ions and therefore expedited where the water contains dissolved acids such as carbonic acid or even organic acids produced during the decomposition of organic matter. The presence of dissolved carbon dioxide will also influence the form of the products of hydrolysis, tending to yield carbonates rather than hydroxides. The hydrolysis of orthoclase felspar is often used to exemplify this type of reaction, and is commonly presented by the equation:

$$K_2Al_2Si_6O_{16} + 3H_2O \rightarrow H_4Al_2Si_2O_9 + 2KOH + 4SiO_2$$
$$\text{orthoclase} \qquad\qquad \text{kaolinite}$$

or, in the presence of carbon dioxide,

$$K_2Al_2Si_6O_{16} + 2H_2O + CO_2 = H_4Al_2Si_2O_9 + K_2CO_3 + 4SiO_2$$

The actual hydrolysis of orthoclase is much more complicated than the equation would suggest and requires a complete disruption of the felspar structure and a rebuilding of the completely different kaolinite lattice. However, the general characteristics of the change are illustrated in that a new mineral species is formed with an associated loss of silica and the production of simple bases.

The idealised hydrolysis of olivine may be quoted as a further instance of this type of reaction:

$$3MgFeSiO_4 + 2H_2O \rightarrow H_4Mg_3Si_2O_9 + SiO_2 + 3FeO$$
$$\text{olivine} \qquad\qquad \text{serpentine}$$

The solvent power of water, providing drainage facilities are present, serves to remove soluble products from the site of reaction, as well as to dissolve out any easily soluble minerals that are originally present in the rocks. If drainage is impeded or insufficient, then soluble constituents accumulate and this situation is reflected in the development of particular soil types which will be discussed more fully in a later chapter.

Hydration of reaction products frequently accompanies hydrolysis, and may take place directly with some mineral species. Examples which can be cited are the hydration of anhydrous calcium sulphate to *gypsum* $CaSO_4.2H_2O$ and the conversion of *haematite* to *limonite* $2Fe_2O_3.3H_2O$. Hydration of more complex primary minerals can also occur as in the formation of *zeolites* from *felspars* and the alteration of the mica minerals, *muscovite* and *biotite*, to clay minerals of the *illite* and *vermiculite* types.

A further chemical factor involved in the alteration of minerals during weathering is oxidation. The most significant change in this connection is the conversion of ferrous iron to ferric iron as occurs during the alteration of *chlorite*, itself an alteration product of several primary minerals, and in the oxidation of FeO, as produced during the weathering of *olivine*, to give FeO.OH (goethite or lepidocrocite), $Fe(OH)_3.nH_2O$ (limonite) or Fe_2O_3 (haematite) according to temperature and moisture conditions. Another important oxidation reaction is the conversion of *iron pyrites* FeS_2 to either oxide or hydroxide with the formation of free sulphuric acid.

Carbon dioxide is important in that the production of carbonates such as *siderite* $FeCO_3$, *magnesite* $MgCO_3$ and *calcite* $CaCO_3$ is made possible, and also for the fact that the solubility of many minerals is increased in its presence, this being especially important in the case of calcium carbonate. The production of alkali carbonates during the breakdown of siliceous minerals provides also a solvent for SiO_2 and thereby a means of translocating silica during the weathering process.

SOIL PARENT MATERIAL

The weathering reactions described above collectively contribute to the heterogeneous weathered complex which eventually becomes incorporated in the soil. But, as has been mentioned, this complex alone does not constitute soil, although it can be considered as providing the soil parent material, which is often regarded as primary if it originates directly from crystalline igneous rocks, and secondary if it is formed from sedimentary rocks which themselves have undergone at least one weathering cycle.

Whether primary or secondary, it does not follow that the soil parent material in any one locality has been derived directly from the rock over which it lies, and it is convenient to differentiate between the modes of origin of the weathered rock products. In hilly regions the unconsolidated residues on the surface of the land tend gradually to work down the slopes under the influence of gravity, a process known as *solifluction*. The movement may be induced by frost and rain action or by the stirring of the surface as occurs during cultivation, or it may be due to an actual downward flow of the loose material under its own weight. The accumulated material at the foot of such slopes is known as *colluvium* and soil developing on it as *colluvial* soil.

Wind is also responsible for the transport and redeposition of parent material, especially in arid climates. One characteristic of wind-borne deposits is the uniformity of grain size and the smooth roundness of individual grains, and two main types of wind-accumulated material can be recognised, viz. dune deposits and loess.

Dunes are usually coarse-grained and siliceous deposits accumulating especially around many coastlines and in desert areas, whereas loess deposits have much smaller individual particles, mainly coming into the category of silt (see page 25), the size range being very narrow. Loess deposits are widespread throughout the European and Asiatic continents. Similar American deposits are often known as adobe, and the brick earths of England are thought by many to be aeolian in origin. The wind also deposits dusts of cosmic or volcanic origin which on land are imperceptible, but which accumulate in significant amounts in the abysmal red clay deposits of the deeper oceans. Transportation of parent material is also accomplished via the agency of ice movement, large areas in Europe and North America being covered by sediment left behind following the retreat of the ice masses of a glacial epoch. Such residues are known as *glacial till* or *boulder clay*. Running water is a very important agent in the transport of weathered rock products, bringing down suspended solids in rivers and streams and depositing them when the speed of the water is sufficiently checked. Deposits occurring in valley flats are known as *alluvium* and give rise to alluvial soils; or if the streams flow into enclosed lakes the suspended matter is deposited on the lake bottom, giving *lacustrine* deposits, or occasionally in estuarine areas, where precipitation occurs in the region where the river and tidal sea waters meet, giving a deposit known as *warp*.

Composition Changes in Parent Material

The nature of soils developed from any parent material is influenced among other things by the extent to which it has been weathered, and it is useful to have some index by which to measure this factor. It should be understood at the same time that weathering of mineral matter does not cease when a soil is formed but is in the nature of a dynamic process, which in mature soils may attain a near equilibrium state, but which in younger soils is still an active and formative reaction.

While it is not possible to trace with absolute precision the chemical mechanism involved in weathering and to formulate stoichiometric relationships, it is possible to obtain estimates of changes in relative amounts of constituent elements. A technique for doing this was worked out by Merrill. One constituent element is taken as a standard, choosing one which is least likely to be lost from the system during weathering. If it is assumed that this element is not removed at all during the breakdown and remains a constant factor, the proportional losses or gains of other ingredients can be calculated.

Merrill used ferric oxide or alumina as reference standards and Table 2 gives an example of his calculations as on basic igneous rock. The increased percentages of sesquioxides in the weathered diorite over the unweathered rock are due to the fact that there is a reduction of total mass during the weathering process which gives rise to a relative enrichment of unweathered constituents.

These figures indicate a marked loss of alkali and alkaline earth metals, and an appreciable loss of silica from those rocks in which it is present combined as silicates; this latter loss is reduced where the silica exists partly in the form of quartz, as in the granite, whilst in the limestone, where any silica would be expected to be present as highly insoluble amorphous silica, it is used as the reference compound.

Cobb applied this technique to a wider range of weathered products, studying rock, weathered rock, young soil, intermediate soil and mature soil, derived from both acidic and basic igneous rocks. The same trends of composition change persist through the different ages of the soils, emphasising the point that mineral weathering proceeds along with soil formation and development.

Some indication of an ultimate equilibrium state is shown in the approaching similarity of analysis of mature soils, especially

TABLE 2
ANALYSIS OF FRESH AND DECOMPOSED DIORITE (Merrill)

Constituent	Fresh Diorite	Decomposed Rock	Decomposed Rock Percentage of original rock on assumption of no loss of Al_2O_3	Calculated loss for entire rock	Percentage of each constituent saved	Percentage of each constituent lost
SiO_2	46·75	42·44	29·32	17·43	62·69	37·31
Al_2O_3	17·61	25·51	17·61	0·00	100·0	0·00
Fe_2O_3	16·79	19·20	13·26	3·53	78·97	21·03
CaO	9·46	0·37	0·26	9·20	2·70	97·30
MgO	5·12	0·21	0·15	4·97	2·83	97·17
K_2O	0·55	0·49	0·34	0·21	61·25	38·75
Na_2O	2·56	0·56	0·39	2·17	15·13	84·87
P_2O_5	0·25	0·20	0·14	0·11	44·00	56·00
Ignition	0·92	10·92		0·00	100·0	0·00
Net loss of original material	—	—	38·53	—	—	—
Total	100·01	99·99	100·0	37·62	—	—

when compared with the diverse analysis of the parent materials (Table 3).

It should be noted that the constitutional changes considered above all take place under climatic conditions in which rainfall exceeds evaporation, and in which soluble products of decomposition are removed by leaching from the system. Under more arid conditions the accumulation of soluble salts makes an overall assessment of weathering changes more difficult.

It is evident that weathering of rocks will result in the formation of smaller and smaller particles, and it follows that the extent of

TABLE 3
COMPARISON OF ANALYSES OF CONTRASTING IGNEOUS ROCKS AND MATURE SOILS DERIVED FROM THEM (Cobb)

	Rock		Soil	
	Granite	Basalt	Granite	Basalt
P_2O_5	0·13	0·34	0·08	0·13
K_2O	3·75	1·84	0·54	0·54
CaO	1·50	8·06	0·33	0·46
MgO	1·13	5·99	0·45	0·59
SiO_2	70·47	52·37	46·70	45·94
Al_2O_3	14·58	15·72	27·13	21·29
Fe_2O_3	3·57	9·88	12·20	19·31
Na_2O	2·99	3·28	0·40	0·13
TiO_2	0·61	1·24	1·73	1·92
MnO	0·07	0·15	0·04	0·37

weathering will be most reflected in the constitution and composition of the finer fractions (viz. the silt and clay fractions) of the soils that eventually develop.

Attempts have been made to ascertain the degree of weathering to which soil mineral matter has been subjected by examinations of the silt and particularly the clay.

Robinson and Richardson proposed a possible method of approach suggesting that an approximate assessment of the degree of weathering could be obtained by calculating the ratio of aluminium present in the clay fraction of a soil to the total aluminium in the soil. This technique assumes that the clay fraction is exclusively produced by the weathering of alumino silicate minerals, and that aluminium occurs only in this type of mineral. The degree of weathering approaches 100 per cent as the ratio approaches unity.

More recently Jackson *et al.* have endeavoured to trace the

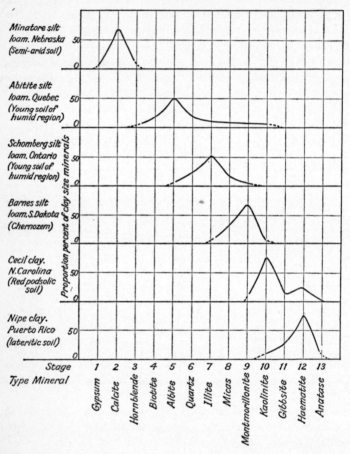

FIG. 2.1   Stages of weathering of clay size minerals as illustrated by clay mineral composition curves of different soils (after Jackson *et al.*).

course of weathering of the finer mineral particles, and have suggested a 'weathering sequence' of the clay size minerals occurring in soils. The sequence is based upon the different susceptibilities of minerals to weathering influences. It involves thirteen stages, each characterised by the dominant presence of certain type minerals;

these each representing one stage of weathering and in increasing order of maturity are: (1) Gypsum, (2) Calcite, (3) Hornblende, (4) Biotite, (5) Albite, (6) Quartz, (7) Illite, (8) Hydrous-mica, (9) Montmorillonite, (10) Kaolinite, (11) Gibbsite, (12) Haematite, and (13) Anatase. There is, of course, an overlap of stages in any particular weathered soil, but usually one stage is dominant. Figure 2.1 (after Jackson *et al.*) gives the clay mineral distribution curves of a range of soils at various stages of weathering.

In soils which are at an advanced state of weathering minerals from stages 1 and 2 are often found following secondary deposition of these more soluble and hence more mobile mineral types, and the presence of any stage of primary minerals introduces modifications in the general interpretation of the sequence.

The assessment of weathering relationships between soils and parent material is made difficult by the vertical variations occurring in soils between surface and deeper levels. The causes and nature of these soil variations will be dealt with in the section of the book dealing with the soil profile, but it is relevant at this stage to point out the importance of ensuring that the materials under comparative study are in fact related. One method of establishing the common origin of soil horizons has been worked out by Marshall involving a quantitative study of certain heavy minerals referred to on page 14, resistant to physical and chemical breakdown. In materials derived *in situ* from a common source the relative proportions of highly resistant minerals, e.g. zircon, tourmaline, anatase and rutile, should be constant, and the particle size distribution of the resistant mineral fraction, should remain the same.

3   The physical nature of soil

The mineral constituents of the soil are represented by particles of widely varying size, shape and chemical composition, and the most obvious differences between individual soils are manifest in the rela-

TABLE 4

SIZE GRADES WITHIN THE FINE EARTH SOIL FRACTION

Atterberg grade	Coarse sand	Fine Sand	Silt	Clay
	2 mm 0·2 mm	0·02 mm	0·002 mm Log. Size Scale	
American grade	V.C. C.M. F. V.F. Gravel - - - - - Sand - - - -		Silt	Clay
Average number particles per g	350	350,000	3×10^8	5×10^{11}
Average surface area per g	40 cm²	400 cm²	4,000 cm²	60,000 cm²
Typical mineralogical make-up	Quartz Felspars Rock fragments	Quartz Felspars Ferro-magnesium minerals	Quartz Felspars Ferro-magnesium minerals Heavy minerals	Quartz Felspars Secondary clay minerals
General characteristics	Loose grained, non-sticky, air in pore space of moist sample, visible to naked eye.	Loose, grained, non-sticky, no air in pore space of moist sample, visible to naked eye.	Smooth and flour-like, non-cohesive, micro-scopic.	Sticky and plastic, microscopic to sub-micro-scopic, exhibit Brownian movement.

(V.C. = very coarse, C. = coarse, M. = medium, F. = fine, V.F. = very fine)

tive proportions of the various size ranges encompassed by their constituent particles. Indeed, most utilitarian classifications of soils are based on such differences in mechanical composition.

Under natural conditions the soil particles do not necessarily exist separately, the smaller ones especially occurring in aggregated units, and in the study of the soil in the field the properties of these aggregates assume considerable importance. It is, however, a convenience to consider the composition of soil first in terms of prime particles, a convenience justified by their intrinsic and fundamental nature as opposed to the impermanent nature of soil aggregates, and a considerable volume of information is available concerning soil make-up in these terms.

Different categories of particles, defined by upper and lower size limits, are recognized, the ranges being, by necessity, arbitrary in character, and much confusion has existed in the past because of lack of standardisation by different workers. In this connection it is usual to accept as soil material only that fraction of a bulk sample which, after air drying and careful crushing to separate loosely held crumb components, will pass through a 2 mm round-hole sieve. Samples of soil obtained in this way are referred to as 'fine earth' samples. Mineral fragments too large to pass through this size sieve are known as gravel, pebbles, cobbles and boulders in increasing order of magnitude.

Three main groupings of soil particles are in common use, namely sand, silt and clay, the groups being variously subdivided according to requirements. The sand and silt fractions mainly consist of primary and secondary minerals derived directly from the parent material, the more refractory minerals such as quartz predominating in the coarser sand, and softer minerals such as felspar and muscovite mica, together with secondary minerals, making up the bulk of the silt fractions; the clay fraction contains few of the ordinary rock minerals, consisting largely of new mineral species collectively known as clay minerals and having distinctive chemical and physical properties.

The size fractions usually used in this country in the mechanical analysis of fine earth samples are those adopted by the International Society of Soil Science in 1927, viz.:

Fraction			Size limits expressed as particle diameters
Coarse Sand	.	.	2·0–0·2 mm
Fine Sand	.	.	0·2–0·02 mm
Silt	.	.	0·02–0·002 mm
Clay	.	.	<0·002 mm

These size limits are based on a grading system suggested by Atterberg and the method by which they are usually separated, which will be discussed below, is based upon the method formulated by a sub-committee of the Agricultural Education Association of Great Britain.

B

A more detailed classification is often adopted, particularly in America, listing seven grades:

Fraction	Size limits expressed as particle diameters
Fine Gravel . .	2·0–1·0 mm
Coarse Sand . .	1·0–0·5 mm
Medium Sand . .	0·5–0·25 mm
Fine Sand . .	0·25–0·10 mm
Very Fine Sand .	0·10–0·05 mm
Silt . . .	0·05–0·002 mm
Clay . . .	<0·002 mm

It is doubtful whether, except in special cases, the subdividing of the sand fraction into so many different categories is of much value or importance, and the International classification has the merit of simplicity, and the equal logarithmic intervals between the successive size limits facilitates the graphic representation of analytical results.

Mechanical analysis refers to the gradation of prime mineral particles, and it is therefore necessary to ensure complete breakdown of particle aggregates prior to an estimation of the fractions. The preliminary treatment of the sample is an important aspect of the analysis and must be done thoroughly if accuracy and reproducibility of results is to be attained. The necessity for complete removal of organic matter in the soil was shown by Robinson, who pointed out that adequate dispersion is not obtained in its presence, and the use of hydrogen peroxide as a reagent to oxidise the organic complex can be recommended as it is free from the objection of introducing any serious mineral decomposition, and moreover, the products of the reaction, carbon dioxide and water, are without effect on the estimation. An alternative treatment was proposed by Troell in which sodium hypobromite is used, excess reagent being decomposed after oxidation is complete by the addition of dilute ammonium hydroxide.

Destruction of organic matter does not in itself bring about a removal of all the factors causing aggregation of the finer particles of soil. To ensure complete dispersibility it is important that the clay fraction should be freed from polybasic cations which cause flocculation of clay suspensions. This may be done by leaching the organic matter-free soil with dilute acid followed by washing to remove residual acid, the final dispersion being done mechanically by shaking with a dilute alkali solution; ammonia, sodium carbonate or sodium hydroxide have been suggested by different workers at different times.

Sodium metaphosphate is commonly used as a dispersing agent in the United States.

There have been, since interest in mechanical analysis of soil developed, a number of methods suggested for the separation of the fractions. Dry or wet sieving for the quantitative assessment of the size grades involved in soil work is unsuitable except for the coarser sand fractions, and the basis of all methods of separation is a differential rate of movement through a viscous medium. If a heterogeneous mixture of solid particles is suspended in a liquid of lower density and the suspension allowed to stand, the particles will begin to settle under the influence of gravity. The individual particle velocities will increase until the frictional resistance between particle and liquid equals the gravitational force, at which point the acceleration will be reduced to zero and the particle will settle at a uniform speed. The value of this constant velocity is governed by particle size together with other factors, namely the difference in density between the solid and the liquid, the viscosity of the liquid, and the acceleration due to gravity. For rigid spherical particles, G. G. Stokes derived an equation relating these factors, particle size being measured in terms of

radius. This equation, known as Stokes' Law, is $v = \dfrac{2}{9} g r^2 \dfrac{\sigma - \varrho}{\eta}$,

where v represents the steady falling velocity, g the acceleration due to gravity, r the particle radius, σ and ϱ the densities of particle and liquid respectively, and η the viscosity of the liquid. The application of Stokes' Law to suspensions of soil particles is subject to one or two limitations and demands a number of precautions and assumptions. The limitations are imposed when the particles are so small in relation to the liquid molecules that Brownian bombardment becomes appreciable and interferes with the rate of fall, and also at an upper size limit when the falling velocity is such that a portion of the liquid in front of the particle is carried down with it. Values for this upper 'critical radius' have been calculated by Arnold and for soil particles through water this critical upper radius is approximately 0·24 mm.

The precautions to be observed are based on the fact that the law applies to free-falling particles which are not influenced by the proximity to the walls of the containing vessel. This means that a vessel of sufficient size must be used for the sedimentation, although no problem is usually encountered in this connection. A more important influence is the interaction between neighbouring particles, the concentration of the suspension being limited by such possible interference. A 2 per cent aqueous suspension is customarily used in actual analyses, and suspensions containing more than 5 per cent by weight will probably be subject to error. A further practical precaution, especially in critical work, must be considered, due to the dependence of the rate of flow on the liquid viscosity. This in the case of water varies rapidly with changes in temperature.

Temperature °C.	5	10	15	20	25	30
Viscosity . . .	0·01518	0·01308	0·01140	0·01005	0·00894	0·00801

As velocity is inversely proportional to viscosity the magnitude of this temperature change effect may be seen by the fact that a change in temperature from 15°C. to 25°C. would increase the falling velocity by a factor 1·28.

The application of Stokes' Law to sedimentation of soil particles is not, strictly speaking, valid as the formula was derived for solid spheres, and soil particles, whilst possessing many geometric shapes, rarely if ever exist as perfect spheres. While the coarser sand fractions may be roughly spherical, the silt and clay fractions occur variously in plate-like or rod-like forms, and it has been shown by Kelley and Shaw that the settling velocities of rod-shaped particles are subject to variations which introduce inaccuracies into their separation by sedimentation. These difficulties can be offset by using settling velocity as the index of soil particle size rather than linear measurements of a diameter, and the particular grades are then defined as comprising particles having settling velocities within the range of those of spherical particles of the upper and lower size limits.

A further difficulty in applying the law to soil separations is that of ascribing a definite density to the mineral components. Owing to the heterogeneity of soil material different constituents will have different densities and hence different settling velocities, so that, especially with particles around the grade boundary limits, a certain amount of overlap may occur. This error is not of any great significance in routine mechanical analysis, but becomes significant when the sedimentation technique is applied to the further subdivision of the clay fraction when accurate density determinations must be made (see page 32).

For routine purposes it is customary to accept an average density of 2·65 but for critical work a more precise figure should be determined. The densities of some common soil minerals are given in Table 5.

For a comprehensive and critical account of the many methods that have been proposed for the actual determination of mechanical composition, most of which have been shown to be inaccurate for one reason or another, many being of historical interest only, the reader should consult the works of Keen. More precise work utilises almost exclusively the so called pipette method, developed independently by

TABLE 5
DENSITIES OF SOME COMMON SOIL MINERALS

Quartz	.	.	.	2·50–2·80
Orthoclase	.	.	.	2·39–2·62
Muscovite	.	.	.	2·70–3·10
Haematite	.	.	.	4·50–5·30
Limonite	.	.	.	3·60–4·00
Kaolinite	.	.	.	2·60–2·68
Montmorillonite	.	.	.	2·53–2·74

a number of workers, whilst for routine determinations where large numbers of samples have to be analysed as quickly as possible the hydrometer method of Bouyoucos is often employed. Both methods depend upon the fact that at any given depth in a settling suspension the concentration varies with time, as the coarser fractions settle at a faster rate than the finer. In the pipette method the suspension is sampled at a given depth at precise times calculated to exclude from

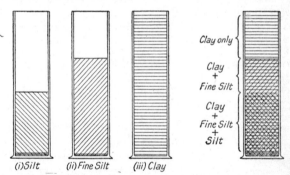

FIG. 3.1   The settling of particles from suspension.

the samples progressively larger size fractions which have wholly settled from the upper layers. Figure 3.1 gives a diagrammatic representation of the principle applied to a suspension of three possible fractions.

The samples are usually taken by means of a pipette and the method is commonly known as the 'pipette method'. The estimation of the percentage of particles belonging to each of the fractions enables a statement of the mechanical composition of a soil to be made with a certain amount of precision. However, the division into (say) four fractions is, for many purposes, not sufficiently detailed. Two soils may each show the same percentage of clay, but one may be largely composed of particles a little less than 0·002 mm while the others may have a preponderance of particles very much smaller. A difference like this, which will have marked effects upon the properties of the soil, is not shown in the mere statement of the percentage of the clay fractions. It is possible to adapt the pipette method of mechanical analysis to further subdivisions, but if carried too far practical difficulties arise.

Ideally, some form of continuous record of the settlement of particles in the suspension is desirable to indicate variations within any particular fraction or fractions. Of the many methods to accomplish this only the technique of Bouyoucos minimises the inherent errors involved. In this technique a hydrometer is used to measure the density of the suspension, and the continuous changes in density can be correlated with the settlement of particles. The difficulties associated with this technique are mainly centred round the fact that the densities recorded are mean figures of a density gradient existing in a volume of suspension corresponding to the submerged length of the hydrometer. Nevertheless, as size distribution curves are usually smooth this average does not lead to serious errors once the hydrometer has been accurately calibrated, and the method has the advantage of simplicity and speed.

Particle Distribution Curves

A continuous representation of mechanical composition can be obtained using the results of a 'pipette method' analysis by constructing a summation curve, plotting the sums of the fractions against the logarithm of the settling velocity, the latter giving, for the International grading scheme, equal increments of the abscissa for each fraction. Figure 3.2 gives examples of summation curves obtained by G. W. Robinson for A, a medium loam soil; B, a clay soil; and C, a sample of powdered slate.

This type of representation of mechanical analysis results, in addition to giving an insight into the continuous nature of the particle size distribution, has the advantage that analyses using different grading standards can be plotted on the same curve, so giving direct comparisons.

As the clay fraction contains the most reactive of the soil mineral constituents it is frequently necessary to determine the size distribution within this fraction and the ordinary sedimentation techniques described above become unsatisfactory for several reasons. The clay particles are of colloidal dimensions and the rate of fall through water is such that analysis would take several weeks to perform, during

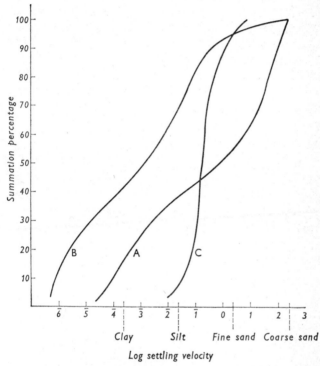

FIG. 3.2   Particle size in terms of log settling velocity (cm/s). A, medium loam soil; B, clay soil; C, powdered slate.

which time the suspension would have to be maintained at a uniform and constant temperature. The method is impracticable for the separation of particles having diameters less than 0·0002 mm (200 nm). However, a method has been described by Marshall whereby particles within the range 2μm to 50 nm can be conveniently separated. The method utilises a centrifuge to increase the forces acting on particles in suspension and so accelerate the sedimentation rate. In

Note: 1μm = 0·001 mm, 1 nm = 0·001μm

the centrifuge method the aim is to allow the different sized particles to settle through a particle-free dispersion medium, all particles being initially held in the same place. Under these circumstances the coarser particles would reach the bottom of the centrifuge tube first, being followed by successively smaller grains. By centrifuging for any given time, the original mixture of particles could be divided quantitatively into fractions, the larger having settled and the smaller being still in suspension. If the particles started from the same line the division between the fractions would be perfectly sharp with no overlap of sizes, and this ideal is nearly reached by Marshall's method in which a thin layer of the clay suspension is carefully placed on top of a column of a more dense dispersion medium, usually a solution of cane sugar or glycerol of known density and viscosity. The necessary times for a particular separation can be derived from the equation

$$t = \tfrac{1}{2}t' + t'' = \frac{1}{3\cdot81r^2N^2}\left(\frac{\eta'\log\frac{x'}{x''}}{2(\varrho-d')} + \frac{\eta''\log\frac{x'''}{x''}}{\varrho-d''}\right)$$

where t' and t'' are respectively the times taken to traverse the suspension layer and the dispersion layer, η' and η'' the viscosities, and d' and d'' the densities of the respective layers; ϱ is the particle density, r the particle radius, N is the number of revolutions per second and x', x'' and x''' are the boundary dimensions of the layers as shown in Fig. 3.3.

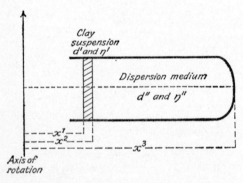

FIG. 3.3   The centrifugal separation of clay size fractions.

A certain amount of overlap in the fractions is inevitable, but if the thickness of the suspension layer $(x'' - x')$ is kept small in comparison with the dispersion layer $(x''' - x'')$, the use of $\tfrac{1}{2}t' + t''$ for a clay with a smooth distribution curve reduces any error to a minimum. Later, Whiteside and Marshall extended the range of this 'two layer' method to the quantitative separation of particles down to 10nm diameter using a Sharples supercentrifuge.

The mechanical composition defines the mineral skeleton of the soil, which in turn confers the major textural characteristics associated with it. Assessment of soil texture is an important feature of practical field work, and differences in texture figure prominently in soil description and classification. The recognition of textural grades in the field is, in practice, based on sensory factors, although ideally it should be related to physical measurements of the mechanical distribution of particles. Field textures are, however, modified by factors other than mineral particles, such as the amount and physical nature of the soil organic matter, the state of aggregation of the silt and clay size minerals, and the nature of the exchangeable cations associated with the clays, although, in the main a fair correlation can be obtained between texture and mechanical analysis.

To define textural designations the composition of a soil is frequently represented as a point with co-ordinates related to the sides of an equilateral triangle in which the three equivalent heights are graduated in terms of percentages of clay, silt and sand, the latter including all particle grades with sizes larger than the upper silt limit. This form of mechanical analysis representation was suggested by Davis and Bennett in America and has been widely used in the U.S.A. In 1945 areas within the triangle were re-defined with appropriate textural connotations (Fig. 3.4) to fit the fraction sizes in common usage there, namely clay 0·005 mm, silt 0·05 to 0·005 mm and sand 0·05 to 2 mm effective diameters.

This arrangement has been modified (Fig. 3.5) to fit in with the International fraction sizes and to accommodate the influence of changing proportions of sand and silt upon the textural properties of the clay.

In order to give a semiquantitative meaning to the terms used to define textural classes limits to their composition have been defined and these are given in Table 6.

Indirect methods of assessing texture with the aim of assigning a definite texture index have been proposed—for instance, by Hardy, who suggested an index of texture based upon moisture content at the point of stickiness and a function of the sand fraction, the former giving a measure of the colloidal content and the latter the modifying influence of coarser material. His expression was I.T. (Index of

TABLE 6
LIMITING VALUES FOR SOIL TEXTURE CLASSES

Sand	Material containing at least 85% sand, provided that the percentage of silt plus 1·5 times the percentage of clay shall not exceed 15.
Loamy sand	Material containing not more than 90% nor less than 70% sand, together with a percentage of silt plus 1·5 times the percentage of clay not less than 15 at the upper sand limit, or a percentage of clay to not exceed 30 at the lower sand limit.
Sandy loam	Material containing either less than 20% of clay with the percentage of silt plus twice the percentage of clay exceeding 30, or having between 43 and 52% of sand with less than 7% of clay and less than 50% of silt.
Loam	Material containing between 7 and 27% of clay, 28 to 50% of silt with less than 52% of sand.
Silt loam	Material containing either more than 50% of silt together with between 12 and 27% of clay, or which has between 50 and 80% of silt with less than 12% of clay.
Silt	Material that contains more than 80% of silt with less than 12% of clay.
Sandy clay loam	Material containing between 20 and 35% of clay, with less than 28% of silt and more than 45% of sand.
Clay loam	Material that contains between 27 and 40% of clay and between 20 and 45% of sand.
Silty clay loam	Material that contains between 27 and 40% of clay and less than 20% of sand.
Sandy clay	Material that contains 35% or more of clay together with 45% or more of sand.
Silty clay	Material that contains 40% or more of clay together with 40% or more of silt.
Clay	Material that contains more than 40% of clay together with less than 45% of sand and less than 40% of silt.

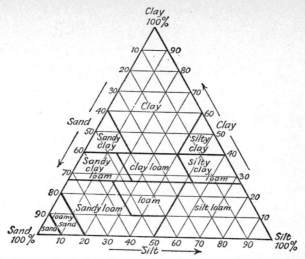

FIG. 3.4 Textural designations according to Mechanical Analysis (American Scale).

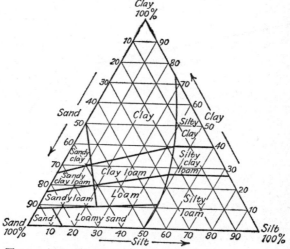

FIG. 3.5 Textural designations as modified by Marshall (Australian CSIR).

Texture) $= P - S/5$, P being the appropriate moisture content and S the percentage of sand, and he obtained values ranging from 10 to 60 with soils of increasing clay content. Others have endeavoured to measure the total surface area of the particles which, being universally related to size, gives some measure of mechanical composition. However, such methods have not proved to be of very great value to field workers.

The Arrangement and Aggregation of the Particles

In the field there are two kinds of ways in which the arrangement of soil particles may be affected. By various cultivation practices the system of particles may be compressed or loosened and the relative proportions between the total volume occupied by solid particles and the pore space varied. The two extreme arrangements for a system of spherical particles of equal size are shown in Fig. 3.6. In the arrangement A the pore space will be approximately 50 per cent of the gross volume, and in the arrangement B it will be approximately 25 per

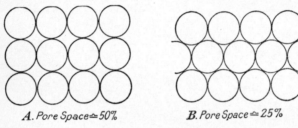

A. Pore Space ≈ 50%   *B. Pore Space ≈ 25%*

FIG. 3.6　Arrangements of spherical particles.

cent. A soil system, however, is different from such a system as this in at least three ways. First the particles are not spherical, second they are not all equal in size, and third the particles are more or less bunched together. A little consideration will show that the formation of bunches or crumbs, will theoretically permit a larger maximum pore space than 50 per cent, and that an assemblage of varied size particles will theoretically allow a minimum pore space below 25 per cent, and that the irregular shape of the particles may influence it either way. As a practical fact, however, the pore space of soils rarely exceeds 50 per cent and rarely falls below 25 per cent. Clay soils usually have the largest percentages of pore space, low figures being associated with sands.

In addition to the purely mechanical ways in which the arrangements of soil particles may be altered, there are other influences which bring about the formation or deformation of compound particles. Alternate freezing and thawing tends to cause particles to aggregate— a fact of considerable importance, particularly on heavy soils. The heat of the sun acts in a similar way, and so it comes about that whereas a farmer in the climate of Great Britain exposes his plough furrow to winter frosts, a farmer in tropical climates exposes it to the sun with very similar results. In both cases the increased tendency for the particles to aggregate is probably due to the dehydration of the colloids.

The effect of electrolytes on the flocculation or aggregation of soil particles is sometimes very marked. In farming practice lime is frequently used on certain types of heavy clay soil to cause the particles to flocculate and to produce what is in effect a lesser number of large particles, thereby rendering the soil lighter. Marked effects of lime on particle arrangement do not however last indefinitely.

In the field the primary soil particles are always to some extent at least combined or arranged in some form of aggregate or secondary particles. Where the aggregation has been contrived by some artificial process the units are referred to as soil crumbs or clods but where the aggregates occur as natural units within the soil profile they are known as 'peds'. These secondary units are arranged occasionally in such a manner as to give a distinct structural pattern within the profile.

It is the extent and kind of aggregation which determines the 'structure' of soil, a feature of great importance to the practical agriculturist who is concerned with tilth production and the maintenance of desirable physical conditions for plant growth.

In order to describe the degree of structure development, four grades of soil structure have been formulated on the basis of inter and intraaggregate adhesion and the cohesion or stability of the units within the profile. They are graded 0 to 3 as follows:

0 Structureless; no observable aggregation or ped formation. If the soil mass is coherent it is *massive*, if non-coherent, *single grain*.
1 Weak—poorly formed indistinct peds.
2 Moderate—well formed, distinctly shaped peds, but not distinct in undisturbed soil.
3 Strong—durable well formed peds that separate into recognisable units when the soil is disturbed.

Schemes of descriptive classifications of structural units have been proposed by the American Soil Survey Association.

These subjective systems follow similar patterns in grouping together structures of similar shapes, with subdivisions according to perfection of geometric definition and size. Clarke, for example, defines five main groups: Cubic, Prismatic, Laminated, Polyhedral, and Single Grain. The cubic group is further subdivided to cover cubic, cloddy, nutty and crumb structures, each in turn having a range of size grades; the prismatic group comprises prisms and columns; the laminated structure, includes plate-like forms and scales; and the polyhedral group includes tetrahedral and polyhedral units of various sizes.

Other workers have made assessments of soil structure characteristics based purely on space measurements or estimations of surface areas for less consolidated soils or on the types of cracking that occur in the more compact and consolidated silts and clays. These, together with the influence of pore space on fertility and moisture status, will be discussed in Chapter 11.

The nature of individual structural units has been investigated by W. Kubiena who, by the use of direct microscopic examination of thin films, has described the nature of different aggregates and the cementing materials holding individual particles together.

He defines six types of aggregates:

1. Loose aggregates of particles, each with a colloidal coating (chlamydomorphous).
2. Aggregates of particles, each with a colloidal coating, bound by colloidal bridges giving a porous system (Plectoamitic).
3. Aggregates of individual mineral particles entirely embedded in colloidal coatings (Porphyropeptic).
4. Undeveloped aggregates of uncoated minerals associated with loosely attached, friable organic colloid (Agglomaratic).
5. Uncoated minerals bound by bridges of friable colloidal material (Intertextic).
6. Massive forms of uncoated minerals split by hair cracks and easily broken down (Porphyropeptic).

The mechanical analysis of soil aggregates can be done by dry sieving for the large units, or wet sieving using a bank of sieves immersed in water, or by an elutriation technique. The wet methods of analysis enable determinations of water-stable aggregates to be made, although there are many difficulties associated with interpreting the results due to vagaries of water stability associated with the treatment of the sample both before, and during, the estimation.

4   Soil clay

Of soil mineral matter, the nature and properties of the finer fractions are by far the most interesting and important, both intrinsically, because of their complexity, and practically, because of their modifying influences on the general properties of the soil in the field.

It was observed over a century ago that soils could participate in 'double decomposition' reactions with solutions of simple salts, exchanging cations with such solutions in stoichiometric proportions, and that these reactions had their origin, as far as the mineral components of the soil are concerned, in certain active silicates associated with the finer fractions of the soil.

The nature of clays generally has long been studied, and for very many years little progress was made in categorising different types of clay although differences, such as base exchange capacity, were appreciated. The difficulties centred around the extreme fineness of the particles, all clays being classed together as colloidal substances and generally assumed to be amorphous in character, although evidence of crystallinity in the clay *kaolin* was recognised. This conception of clay as being composed of colloidal amorphous silicates persisted for many years. A further obstacle hampering progress in the study of clay minerals in soils was the difficulty in obtaining samples of sufficient purity for critical study. In soils the clay is arbitrarily classified as consisting of particles having equivalent diameters of less than 0·002 mm, but this fraction frequently contains non-clay particles, especially in the coarser ranges. Primary minerals such as quartz, felspar, mica, calcite and pyrites may occur in sizes down to $0·5 \mu m$ (0·0005 mm), and to be reasonably certain of excluding all primary rock material only fractions smaller than this should be considered.

The present-day appreciation of soil clays can be said to begin with the recognition by Ross and Shannon in 1926 that a number of highly colloidal clays were essentially crystalline, and they were able to distinguish quantitative differences between these *clay minerals*. By examining thin sections under a petrological microscope they obtained a high birefringence, indicating a crystalline nature, and by coupling refractive index and chemical analysis were able to suggest a classification for mineral types.

C. E. Marshall in 1930 further established the crystalline character of a Rothamsted clay fraction, and quantitatively measured the birefringence of aqueous suspensions of the fine clay by orientating the particles in an electric field. He also showed that the birefringence varied measurably with the cations associated with this clay in contradistinction to kaolin. Marshall's interpretations of his observations were of far-reaching importance, indicating in the first case that the cations were associated with sites within the crystal, whereas with kaolin they were related to external surfaces only. Contemporary application of X-ray diffraction studies by several workers amply confirmed the crystalline characters, and the use of more refined X-ray diffraction techniques, together with the application of principles proposed by L. Pauling following his study of micaceous minerals, has enabled the actual structures of individual clay minerals, with slight reservations in individual instances, to be elucidated.

THE CLAY MINERALS

Apart from the tiny fragments of primary minerals that exist within the clay fraction, its bulk is composed of secondary minerals which are products of weathering. As there are many different source materials, and as the course of their weathering is influenced to differing extents by temperature, moisture and amount of aeration, there is bound to be considerable variety in these secondary products which are often referred to as the clay minerals. As would be expected from the nature of soil mineral matter, silicates figure prominently amongst them although there are several non silicates in the fraction.

The principal clay minerals to be found in soils are listed in Tables 7 and 8.

Colloidal Properties of Clay

As can be seen from Table 4 the number of particles and their surface area per gramme of clay fraction is disproportionately great. It is in this fact that the importance of clay resides. The properties of a whole mineral soil are determined more by this fraction than by any other group of soil minerals. The features responsible for this are both physical and chemical and merit a more detailed discussion.

In physical terms a large proportion of the fraction is classed as being composed of colloidal particles, which when suspended in water have properties in between those of a true solution and those of an aqueous suspension of an insoluble solid. The term 'colloid' was introduced in 1861 by Thomas Graham to distinguish between solutes that diffused easily through a parchment membrane and those which did not.

TABLE 7
PRINCIPAL SECONDARY MINERALS OF THE CLAY SIZE FRACTION

Amorphous	oxides	silica $SiO_2 . nH_2O$ hydrated iron oxides $Fe(OH)_3 . nH_2O$ alumina $Al_2O_3 . nH_2O$
	silicates	allophane $Al_2O_3 . 2SiO_2 . nH_2O$ hisengerite $Fe_2O_3 . 2SiO_2 . nH_2O$
	phosphates	evansite $Al_3PO_4(OH)_6 . nH_2O$ azovskite $Fe_3PO_4(OH)_6 . nH_2O$
	carbonates	calcite $CaCO_3$ magnesite $MgCO_3$ dolomite $CaCO_3 . MgCO_3$ siderite $Fe CO_3$
Crystalline	oxides	goethite $\alpha FeO . OH$ lepidocrocite $\gamma FeO . OH$ maghemite γFe_2O_3 gibbsite Al_2O_3
	silicates	complex alumino-silicates, grouped according to their structural characteristics.

TABLE 8
PRINCIPAL CRYSTALLINE SILICATE CLAY MINERALS

Chain structures		palygorskite
Layer structures	1 : 1 family	kaolinite nacrite dickite halloysite
	2 : 1 family	montmorillonite beidellite vermiculite mica illite nontronite

In a true solution the dissolved substance is usually of molecular dimension and the individual particles which are randomly and uniformly dispersed cannot be seen against the background of solvent molecules. It is a disperse system. If the dispersed particles become progressively larger relative to the size of the solvent molecules it becomes possible to demonstrate their individual presence by means of an ultramicroscope which identifies the scintillations of reflected light when a beam is focused into the dispersion according to a phenomenon known as the Tyndall effect. The individual particles can remain randomly dispersed throughout the whole solvent. The particle size at which this becomes possible is not precisely fixed but is about one nanometre. As the particle size increases still further a limit is reached at which the random dispersal of particles is destroyed, and under the influence of gravity they separate from the dispersion liquid. The size at this stage is approximately one micrometre. Within these limits the suspension is referred to as a colloidal solution or *sol* and the individual particles in it as *micelles*. The micelles within the suspension also exhibit a continuous, haphazard motion in all directions, the so called Brownian Movement, which is caused by an irregular bombardment by the molecules of the liquid in which they are suspended.

Colloidal solutions fall into two classes which to some extent overlap but which are represented in the extremes by the *lyophobic* colloid on the one hand and the *lyophyllic* colloid on the other. In the first there is no chemical affinity between the disperse phase and the suspension medium. They are usually inorganic and consist of very finely divided solids dispersed within but distinct and separate from the solvent molecules. Lyophyllic colloids which are frequently organic show some affinity between particle and solvent which considerably modifies their behaviour as free flowing suspensions. In those cases where water is the solvent the systems are usually called hydrophobic (water hating) or hydrophyllic (water loving). Colloidal clay particles are somewhat intermediate between the two classes possessing many of the properties associated with true hydrophobic sols, especially when examined as very dilute suspensions in the laboratory, and possessing some of the features of hydrophyllic systems particularly when looked at in high concentrations.

Colloidal dispersions often tend to be unstable and when the solid phase separates it is difficult to redisperse it. This is particularly so with hydrophobic sols. The initial stability is associated with the fact that the particles carry what is effectively an electrical charge, and the electrostatic effects prevent the particles in suspension from colliding one with another. The charge may be positive or negative according to the nature of the particle. It may vary in magnitude according to the nature of the solvent and the presence of other solutes and in some cases may change its sign from positive to negative or vice versa. The effective charge is known as the zeta potential.

At minimum potentials individuals particles may collide within the suspension, and when this happens they invariably stick together. Further collisions lead to the formation of aggregates which may increase in size to the extent of causing them to fall out of suspension. The effect is called flocculation or coagulation. Whether or not soil clays are in the flocculated or non flocculated (peptised) state has a marked effect on the physical properties of the soil particularly in terms of porosity. Figure 3.1 diagrammatically shows the difference between flocculated and peptised states. The ease with which clays form stable *flocs* varies with the nature of the clay itself and it is considerably influenced by the composition of the soil solution and the ions associated with the clay micelles as will be shown later.

Clay is not the only colloidal constituent of soil. A considerable proportion of the organic matter present is also in a colloidal condition although in this case the state is much more akin to the hydrophyllic division. In referring to the colloidal fraction of soil it must be appreciated that both mineral and organic components are involved and that these may interact to introduce deviations from the reactions of each considered separately.

The Non-Crystalline Clay Minerals

A considerable proportion of the clay mineral fraction is composed of amorphous oxides of silicon, iron, and aluminium, in various associations. In some soils, particularly in areas of intense weathering, oxides of iron and aluminium may dominate the clay complex. Much of the colloidal amorphous fraction is adsorbed on to other minerals in the soil to form coatings around the larger particles which assist in a number of cases in the development of soil aggregates by acting as a cementing material binding particles together.

The amorphous oxides, because of their disperse nature are also likely to be in a potentially highly reactive state, and iron oxide in particular appears to play an important part in many pedological processes. This oxide, together with aluminium oxide, forms a medium for phosphate adsorption and so can influence the availability of phosphorus to the plant. In the amorphous state it is more readily reduced from the ferric to ferrous state, a change which affects soil profile development in water-logged situations; on ageing, the hydrated iron oxide can crystallise in a variety of waysaccording to the environment, often modifying soil horizons, which may lead to important taxonomic consequences.

Amorphous silicates which also contain aluminium and magnesium are common constituents of the clay fraction and are described as the *allophane* minerals. It is though that the allophanes are in some respect similar to the crystalline silicates and are made up silicon-oxygen tetrahedral units with octahedral magnesium or aluminium associates although in this case they are randomly dispersed so that no formal structural pattern can be elucidated.

The Crystalline Clay Minerals

Calcareous and dolomitic limestones are often parent materials on which soils have developed, and in the clay fraction of such soils crystalline carbonates may occur as inherited minerals. Calcite, magnesite and dolomite can also be found as secondary minerals in some situations where crystallisation from soil water has occurred.

Crystalline iron oxides are commonly found throughout soil profiles, and their presence is often indicative of some particular soil development situation. Goethite and lepidocrocite are indicative of a humid, temperate environment and the latter can only form if a reducing state has been involved, as its open cubic lattice structure can only develop from iron in the ferrous state. Hematite and maghemite are characteristic of warm and dry climates and again the latter crystalline oxide must have had ferrous iron as a precursor.

The silicate clay minerals form a group of considerable variety and importance. They have a great effect on the physical and chemical properties of soil and have a large bearing on the mineral nutrient capacity. The different silicate minerals vary widely in their properties which are linked to their detailed crystal structures. In order to understand the differences and the properties themselves some understanding of these structures is important.

CLAY MINERAL STRUCTURES

Fundamentally the silicate clay minerals can be considered as complex compounds in which silica is combined variously with alumina or magnesia. Cations other than aluminium and magnesium are also involved, both structurally in the crystal framework and in close association with the basic structures.

Crystalline silicates occur in which the silicon and oxygen atoms involved assume various geometric patterns in space, the structures in each case being based on the *silicon-oxygen tetrahedron*; in this, a silicon atom is centrally placed in relation to four equidistant oxygen atoms (Fig. 4.1).

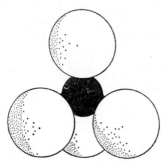

FIG. 4.1   Silicon oxygen tetrahedron.

In the clay minerals generally these tetrahedra form sheet-like structures in which adjacent units, similarly orientated, have their bases in a common plane and in which the individual basal oxygen atoms are shared by adjacent tetrahedra (Fig. 4.2).

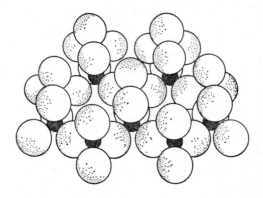

FIG. 4.2   Silicon oxygen tetrahedra in sheet formation.

In this arrangement each silicon atom is in contact with four oxygen atoms, and is described as having a co-ordination number of four, and the basal oxygen atoms which assume a hexagonal pattern may be considered as inactive, each having its two negative valencies

forming covalent links with two separate silicon atoms, whilst the
apical oxygen atoms may be regarded as potentially active, having
one negative valency not absorbed by the structural requirements of
the sheet. The unsatisfied charge on these oxygen atoms may
be neutralised by positive charges which may be furnished by
cations, as in the case of the single finite hexagon of the mineral *beryl*
$Be_3Al_2Si_6O_{18}$, or, in the case of the infinite sheets occurring in the
clay minerals, by incorporation in analogous structures of alumina
or magnesia.

The mineral *gibbsite* $(Al_2(OH)_6)_n$ exemplifies the structures associated with the tetrahedral silica sheets in the clay minerals. Aluminium has a co-ordination number of six, and can assume a geometrically stable structure in which individual aluminium atoms

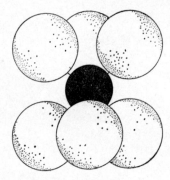

FIG. 4.3   Aluminium hydroxyl octahedron.

assume a position central to six hydroxyl groups arranged octahedrally
(Fig. 4.3).

Two-dimensional extension of this with hydroxyl groups being
shared between adjacent octahedra gives a sheet structure with two
planes of hydroxyl groups in close packing, having aluminium
atoms occupying octahedral sites between them. In order that the
charge status of this sheet arrangement be zero, only two out of every
three octahedral positions between the hydroxyl sheets are occupied
by aluminium, giving what is referred to as a dioctahedral structure.
In the mineral *brucite* $(Mg_3(OH)_6)_n$ the structural features are
identical except for the fact that magnesium atoms occupy all the
octahedral sites available, described as a trioctahedral structure; the
overall charge of this structure is again zero owing to the lower
valency of magnesium (Fig. 4.4).

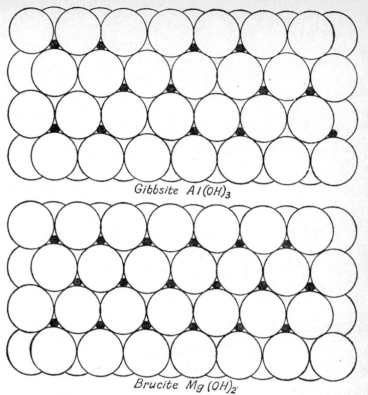

Gibbsite $Al(OH)_3$

Brucite $Mg(OH)_2$

FIG. 4.4   Gibbsite and brucite structures.

The Kaolin or Kandite structure

The amalgamation of the above structures to give the clay minerals can best be illustrated diagrammatically, and for the simplest of the clay mineral types, the kaolin group, can be represented as shown in Fig. 4.5. The arrangement will be seen to involve the substitution of monovalent hydroxyl groups in the octahedral structures by the apical oxygen atoms from the sheets of silicon oxygen tetrahedra. The unsatisfied valency charges of these latter are neutralised by charges within the octahedral framework.

The dimensions of the tetrahedral and octahedral units allow the substitution as illustrated to exist with little strain on the structure, and it can extend uniformly along the *a* and *b* axes. Growth of the crystal along the *c* axis is obtained by a stacking of layers, giving a sequence in which the basal sheet of oxygen atoms of the silica tetrahedra falls adjacent to the hydroxyl layers of aluminium hydroxide octahedra, individual oxygen and hydroxyl groups approaching one another in pairs. Hydrogen bonding can occur at these junctions, giving stability to the crystal along the *c* axis.

Several members of the kaolin or kandite group of clay minerals, which is also referred to as the 1 : 1 family, the structure being

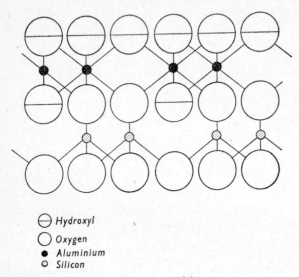

⊖ Hydroxyl
◯ Oxygen
● Aluminium
◎ Silicon

FIG. 4.5   Kaolinite structure (diagrammatic).

based on one silicon tetrahedral layer to one octahedral layer, have been identified, namely *kaolinite*, *nacrite*, *dickite* and *metahalloysite*, with several less well defined materials, the differences being due to variation in the relative position and orientation of the repeating layer units.

Closely related to, but distinct from, kaolinite are the halloysite minerals, in which kaolinite type layers are stacked with intervening layers of water molecules which, according to Hendricks and Jefferson, have a definite configuration. The fully hydrated halloysite can lose water in a conversion to partially hydrated forms.

The Montmorillonite or Smectite Group

The generally accepted structures of clay minerals comprising this group are analogous to the structure of the macrocrystalline mineral pyrophyllite. (The more important members are montmorillonite, beidellite, nontronite, saponite, hectorite and sauconite.) As in the case of the kaolin minerals, the pyrophyllite structure embodies silicon-oxygen tetrahedral units and aluminium hydroxide octahedral units, but in this case two tetrahedral sheets merge into the opposite sides of one alumina sheet, which is the basic reason for this group often being called the 2 : 1 family. This is diagrammatically illustrated in Fig. 4.6. This again is a geometrically stable structure along the a and b axes and development along the c axis is obtained by repetition of the unit layers, but here, repetition along the c axis brings into adjacent positions the planes of oxygen atoms forming the tetrahedral bases. No bonding mechanism exists between the layers which are held together solely by van de Waal's forces.

The distinction between the structures of pyrophyllite and the members of this clay mineral group depends upon the fact that in the latter certain atomic substitutions have occurred, replacing silicon and aluminium from their respective idealised sites, and water and certain other polar molecules can penetrate between the layers, causing an expansion of the crystal along the c axis. These atomic replacements have important repercussions on the properties of the clay minerals and markedly influence their chemical behaviour.

The possibility of atomic replacements taking place in crystals without disturbing the lattice structure depends upon certain rules relating to isomorphism, and where these replacements occur in the clay they are referred to as isomorphous replacements. The dominating factors governing such replacements are the respective sizes of atoms concerned in the changes from the ideal state, and their coordination numbers in the crystallographic sense.

The principal substitutions found in the montmorillonite group are: aluminium substituted for silicon in the tetrahedral positions, and magnesium and iron (ferrous and ferric) for aluminium in the octahedral sites, and others are possible. The nature and extent of isomorphous replacement gives rise to the various members of this group, thus montmorillonite has a proportion of octahedral substitutions of magnesium for aluminium, beidellite has tetrahedral substitutions of aluminium for silicon, and nontronite contains iron in the octahedral cation sites.

The major effect of this isomorphous replacement is in its effect on the charge status of the structure. In the pyrophyllite crystal the positive and negative charges associated with the individual atoms counterbalance. This can be seen from the illustration in Fig. 4.6 below in which the positive charges of the aluminium and silicon atoms exactly equal the negative charges of the oxygen and hydroxyl groups. When, however, isomorphous replacement has occurred the charge balance is changed, and replacement of silicon by aluminium in the silica layers, due to the different valencies, induces one negative charge on the structure for each such replacement. Similarly replacement of aluminium in octahedral positions by ions such as

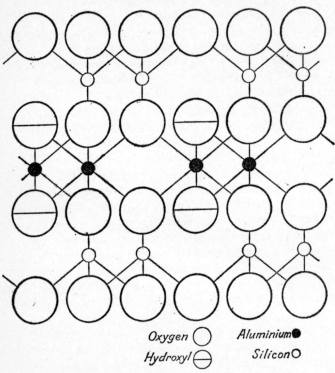

Oxygen ◯   Aluminium ●
Hydroxyl ⊖   Silicon ○

FIG. 4.6   Pyrophyllite structure (diagrammatic).

magnesium or ferrous iron again introduces negative charges inherent in the crystal lattice.

These changes are neutralised normally by low valency cations such as sodium, potassium, calcium and magnesium, which accommodate themselves to the lattice, but, though essential to the system, these ions are not individually part of it, and they constitute the exchangeable cations which can be replaced by simple reactions as mentioned earlier. Isomorphous replacements are not the sole source of charge on the clay mineral crystals. The terminal atoms on the edges of the lattices have charges not fully compensated structurally and these, being normally oxygen atoms or hydroxyl groups, confer an edge charge to be neutralised by external cations in a like manner; in addition under certain conditions dissociation of some structural hydroxyl groups is possible, hydrogen ions dissociating from the structure leaving the micelle with further negative charges.

The Illite Group

A variety of minerals come under this heading and all can be considered as being akin to the macrocrystalline mica minerals. They are also referred to as the hydrous mica group. In certain instances specific names such as *sericite*, *bravaisite* and others have been allotted to particular members, but the range of composition covered by the illite minerals is so wide and the limits of variation so ill-defined that only the general features of the group will be described here. Two main subdivisions occur in which the type structures are those of muscovite and biotite. The structure of muscovite is that of pyrophyllite, in which one quarter of the tetrahedral silicon atoms are replaced by aluminium, giving a negative charge of two units per unit cell. This charge is compensated by two potassium ions which take up structural positions in the hexagonal cavities of the tetrahedral base planes adjacent to the sites of the charge discrepancy. The illite minerals related to muscovite have similar structures except that they are hydrated, and the increase in hydration is accompanied by a decrease in the content of potassium.

Biotite mica differs from muscovite in that the octahedral layer is based on the brucite structure, in which magnesium atoms (see page 46) occupy the octahedral positions. It is trioctahedral as opposed to the dioctahedral minerals involving gibbsite, in which only two of every three such sites are occupied. In biotite a proportion of the magnesium is replaced by ferrous iron. The illite minerals in this subsection show a similar inverse relation between degree of hydration and potassium content.

Unlike the montmorillonite group, the illite clay minerals (both having a 2:1 lattice form, i.e. two tetrahedral to one octahedral layer as opposed to the 1:1 lattice of the kaolin minerals) are non-expanding lattices; although *vermiculite*, which may be considered as the fully hydrated end member of the biotite illites, with no potassium in its structure, does expand, though to a lesser degree than montmorillonite.

The Palygorskite Group

These clay minerals, including attapulgite and sepiolite, differ from the preceding minerals in that they form needle-like crystals and do not have a plate-like habit. They show structural differences whilst still being composed of silica tetrahedral units linked together by octahedral groups. The difference can perhaps be understood by recalling the hexagonal silica sheet structures described earlier (p. 45).

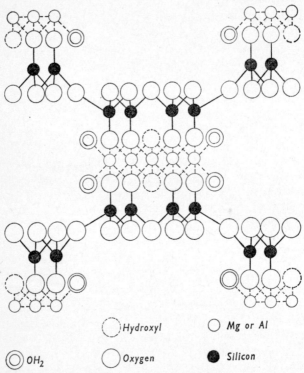

FIG. 4.7 Palygorskite structure (diagrammatic).

In these, the tetrahedra were orientated in similar fashion, and the sheet extended evenly in both *a* and *b* axes. In the palygorskites, similar hexagonal groups of tetrahedra are present but adjacent chains of hexagonal groups have their apices reversed. The apical oxygens of one layer are linked to their counterpart atoms of the next layer by aluminium and/or magnesium in octahedral co-ordination. The structure is illustrated in Fig. 4.7 (after Bradley).

The structure has a system of open channels or pores which contain water molecules and in which exchangeable cations can be accommodated.

Cation Exchange

The property of cation exchange associated with the clay minerals has already been mentioned, and because of its fundamental and practical importance, much research into the kind and amount of exchange taking place between ions of different elements has been attempted. The capacity of different clay minerals for exchange varies widely, and even for a single mineral there is no fixed exchange capacity characteristic of the mineral. The ranges of cation exchange capacities for different clay minerals are given in Table 9.

TABLE 9
CATION EXCHANGE CAPACITIES AT pH 7 OF CLAY MINERALS, EXPRESSED IN MILLI-EQUIVALENTS PER 100 GRAMS OF CLAY

Kaolinite	3–15
Halloysite (fully hydrated) . . .	40–50
Illite.	10–14
Montmorillonite	80–150
Vermiculite	100–150
Palygorskite	20–30

As has been explained above, it is now considered that the cations taking part in exchange reactions are those which, whilst being an integral part of the clay crystals, are not structural components, being present to furnish the balancing charges necessary to ensure electrical neutrality of the crystal micelle. According to Pauling's principle of local satisfaction of electrostatic forces in crystal structures, these balancing ions should take up positions as near to the charge discrepancy sites as possible. Thus in the kaolinite minerals in which there is negligible isomorphous replacement charge, and if only broken bond charges at the crystal edges exist, these will be neutralised by balancing cations which will tend to congregate around the edges of the crystal. In the montmorillonite type, where appreciable isomorphous replacement may have occurred, there will be balancing ions at the crystal edges and also along the crystal sheet planes, both externally and internally where the expanding lattice allows freedom of entry, the ions approaching the unbalanced replacement sites as closely as possible. In all mineral types there is also the ionisation of hydrogen ions from exposed hydroxyl groups, which will contribute to the exchangeable cation content provided the pH of the environment is sufficiently high to permit of such dissociation, or the dissociation of hydroxyl ions from the structure of the pH is sufficiently low.

The exchange capacity of a clay mineral is therefore the consequence of four types of charge:

(a) edge oxygen atoms or hydroxyl radicals incompletely neutralised by the crystal structure;
(b) charges induced as a result of isomorphous replacements with the structure;
(c) dissociation of hydrogen ions at high pH levels;
(d) dissociation of hydroxyl ions at low pH levels.

The dependence of charge upon pH means that the exchange capacity of a clay mineral is not constant and must be measured for purposes of comparison under standard conditions. The major contribution to exchange capacity is, however, made by 'edge' and by isomorphous replacement charges.

While any kind of cation in solution can take part in exchange reactions with clay minerals, all cations do not exchange with the same ease, and some cations, when in association with the mineral structures, are more difficult to displace than others. There are no hard-and-fast rules which govern the behaviour of cations in this respect, but the rules suggested by Wiegner still afford the best generalisations to describe the relative behaviour of different cations. According to these rules, for ions of equal valency, those which have the highest degree of hydration have the least energy of replacement and are most easily displaced when already present in the clay, and the ease of replacement is greater the lower the valency. Thus it can be generally expected that ease of entry of some common cations will be $Li < Na < K < Mg < Ca < Ba < Al$. Hydrogen ions are somewhat difficult to place in such a scheme, and the order is frequently disturbed by concentration variations and by idiosyncrasies of particular mineral species.

5   Soil organic matter

Organic matter is an essential and characteristic constituent of soils. It is present in a variety of forms ranging from undecomposed plant and animal remains to amorphous dark coloured substances which are products of complex decomposition actions.

Plant residues ranging from more or less succulent leaves to highly lignified woody residues constitute the principle sources of this organic matter and microbiological processes cause the chemical and physical changes involved. The processes of change bring about the production of simple end products of decomposition CO_2, H_2O, NH_3, NO_3, SO_4, PO_4, and at the same time lead to the formation of organic residues of a complex kind which have considerable stability and are resistant to further change. In so far as these processes can be regarded as separate the former is referred to as 'mineralisation' and the latter as 'humification'. The relative effectiveness of mineralisation and humification depends upon conditions of environment as well as the nature of the organic matter. Under agricultural conditions in a temperate climate it has been shown that up to 70 per cent of an organic dressing disappeared within 2 years (mineralisation) while the residue disappeared at the rate of about $1\frac{1}{2}$–2 per cent per annum. The more conditions are conducive to microbiological activity the greater is the mineralisation.

The products of humification, however, are important constituents of soil, but it is doubtful if one can claim the existence of compounds or complexes that have permanent existence within the soil in a way that soil minerals can be regarded as permanent. It is fairly certain, however, that soils contain a stable reserve or reservoir of compounds or complexes of individually characteristic nature which constitutes the humus fraction, and most detailed studies of soil organic matter are aimed at resolving the nature of the equilibrium complexes associated with different soil situations.

In organic rich surface horizons, several stages of organic residues can be recognised:

1. Fresh or near fresh debris, little attacked and having recognisable physical structures.
2. Intermediate products of decomposition including residues resulting from the differential disappearances of compounds from fresh debris.

3. Colloidal complexes synthesised by microbial action (and including organisms themselves).

The main difficulty in the study of soil organic matter is the partitioning and separating into different fractions material which in fact comprises a continuous state of matter. It is inevitable that any system of fractionation must be empirical, and to give the most useful information it is important that the techniques employed do not modify the materials being studied. It can be argued that for any particular soil it is only the more stable humified material that can be considered as being characteristic of that soil, and that this fraction is the one whose properties relate to the soil. The non- and only partially-humified material is ephemeral as far as the soil is concerned and should only be considered as having a temporary or peripheral influence. In agronomic terms, however, these latter fractions can have a real bearing on soil fertility, influencing ease of cultivation, physical properties, and the continuous provision of available nutrients through mineralisation. The relative proportions of humified and non-humified material is often characteristic, relating to the rate of turnover within the biological cycle, and some assessment of degree of humification in many instances gives useful data in the study of soil genetics.

FORMATION OF HUMUS

The bulk of seasonal accumulation is deposited on the surface of the soil as leaf litter or dead plant material but there is also a by no means negligible underground contribution from root systems together with a miscellany of faunal residues. The fresh organic matter may, as has already been suggested, be either (*a*) oxidatively decomposed or mineralised and converted into simple end products, (*b*) physically incorporated into the solum, or (*c*) remain on the surface as an accumulating layer.

The first condition requires a highly active microbial flora, and this is usually associated with an aerobic and natural or slightly alkaline environment. It leads to considerable modification of the organic matter and the production of breakdown products having little resemblance to the original substance of the fresh material.

The microbiological population of normal soils is high, involving a variety of organisms. Bacteria and Actinomycetes are present in a variety of genera and species, most of which are heterotrophic, using preformed organic carbon and nitrogen compounds as nutritional substrates although autotrophic organisms are also present which can synthesise their body substances from simple inorganic compounds. In number they may range from less than one million to over three thousand million organisms per gram of soil. In addition a varying population of fungi, mainly Phycomycetes, inhabit the soil, and probably these are all heterotrophic although their food requirements may vary widely. Soil algae which elaborate carbohydrates photosynthetically from atmospheric carbon dioxide, and protozoa, which in the main utilise bacteria as a food source, are also present in the soil micropopulation.

The various heterotrophic organisms rely for continued growth and development on organic sources of carbon from which their body substances can be elaborated, and can use compounds ranging from simple sugars to complex polysaccharides, the breakdown of which provides the energy necessary for resynthesis. The nitrogen requirements of the organisms are met usually by amino nitrogen derived from proteins or other organic sources, although many organisms can utilise nitrogen from inorganic compounds and a few species can use gaseous nitrogen. Organisms such as these are responsible for major decompositions of fresh organic matter in the soil.

The second condition can only occur when there is an active working fauna in the soil and in particular a high worm population which consumes plant debris together with a certain amount of mineral matter, partially digests the organic, and excretes an intimate mixture of the two. Providing the environment is appropriate worm populations can be very high and it has been computed that of the order of 1,000 kilogrammes of worms can inhabit one hectare of land having a soil consumption of something like 90 tonnes per hectare per annum. As soil conditions optimum for worm activity are also favourable for micro-organisms generally the turnover of organic matter can be quite rapid. Mineralisation occurs to a marked degree and the residual products of humification remain in intimate association with the mineral fraction and, in particular, the clay minerals. This gives rise to a humus form described as 'mull' humus in which organo-mineral complexes account for the greater proportion of the organic matter.

Earthworms can only thrive under certain conditions and their universal distribution is limited. By and large they are intolerant of drought and frost and are not therefore found to any great extent in sandy soils or thin soils overlaying solid rock; they prefer an aerated

c

habitat and are therefore rare in very heavy or undrained soils: they also have a continuous requirement for calcium and do not thrive in highly acid soils. These restrictions also apply to many micro-organisms and therefore when such conditions prevail organic matter remains on the surface of the soils in a relatively unchanged condi-tion and often quite sharply separated from the mineral soil, as referred to in the third condition above.

Where litter accumulates in layers of significant thickness it is often possible to recognise three distinct and separate phases. Upper-most is a layer of practically unchanged residues often of leaf, stem and bark which together constitute the 'O' layer of the soil profile. This gives way to a much more comminuted version, partly decom-posed and broken down by insect larvae and smaller fauna such as mites, and although an element of change has occurred it is still possible to pick out pieces whose morphology is sufficiently recognis-able to indicate their origin; this is the 'F' (fermentation) layer. This in time gives way to a further stage of breakdown where all the morphological features have disappeared, a much darker colour has developed and strong evidence of high fungal activity can often be observed by the ramification of hyphae through the layer. This is the 'H' (humification) layer and this form of organic matter is referred to as 'mor' humus. Between 'mull' and 'mor' humus forms there are many intermediate grades, the recognition of which often helps to establish soil types and to determine patterns of soil development as will be discussed later.

The form of humus is not entirely determined by microbiological agencies. The nature of the original organic matter plays an import-ant part. In general grass and deciduous litters tend to mineralise fairly easily and give rise to the more decomposed forms of humus whereas heath and coniferous residues with their more highly lignified nature decompose only slowly and tend to form mor humus. Amongst the common forest litters ease of mineralisation decreases in the order: elm, ash, birch > oak, beech > pine, spruce.

COMPOSITION OF HUMUS

The fresh organic matter which forms the starting point in the humification process is very variable both in character and rate of production. In chemical terms it can be considered broadly as comprising three main groups of compounds, polysaccharides, pro-teins and lignins, together with incidental amounts of other organic compounds such as waxes, alkaloids, essential oils and pigments.

The Carbohydrate Constituents

The polysaccharides or carbohydrates comprise the simple sugars and starches which are of little importance in soil humus as they are very rapidly metabolised by soil micro-organisms to give carbon dioxide and water, together with cellulose, cellulosans and hemi-celluloses which in the plant constitute the encrusting and skeletal tissues. Cellulose and cellulosans are compounds formed by the condensation of hexose and pentose units which include glucose, mannose and xylose.

These compounds would appear to form readily available energy sources for micro-organisms and disappear quite rapidly, cellulose in particular being quickly decomposed. The hemicelluloses are chains of uronic acid units which are principally glucuronic acid, and these are broken down quite effectively by certain soil fungi.

However, carbohydrates are also produced by bacteria and fungi which are again mainly polyuronides, so that within the soil organic matter there will always be a proportion of polyuronides present. In fact these constitute the main carbohydrate fraction of humus. A figure of between 10 and 15 per cent of the organic carbon in the surface horizons of soils is accounted for by such carbohydrates and the proportion tends to increase with depth.

Another polysaccharide-like complex is also to be found in soils. This is chitin which is a condensation product of glucosamine. This is found as an exoskeleton in worms, bacteria and crustaceae and contributes both to the carbohydrate and the nitrogen of humus.

In broad terms then, the carbohydrate fraction of a soil will consist of a number of products ranging from cellulose which is entirely of plant origin to bacterial polysaccharides which have been synthesised within the soil during the humification process.

It is arguable whether the carbohydrates present in soil constitute an integral part of the stable humus complex, and if they do not they should be considered as non-humic material. They are, however, invariably precipitated with humic acid, and can be separated, at least partially, by using solvents such as pyridine or alcoholic sodium hydroxide, in which they are insoluble.

Nitrogen Compounds

The nitrogen content of soils ranges from extreme values of as low as 0·05 to upwards of 2 per cent. Whilst some of the total nitrogen is inorganic, the overwhelming proportion is to be found associated with organic matter and, as the nitrogen in humus constitutes the reservoir from which plants are eventually nourished, some knowledge of how it is present is likely to be useful.

Studies of the types of organic nitrogen compounds present are usually based on recognising the relatively simple compounds that are released when the soil organic matter is treated with hot acids. By such methods between 20 and 50 per cent of the nitrogen can be shown to be in the form of amino acid complexes. This is perhaps to be expected as most of the nitrogen introduced from fresh organic matter is in the form of vegetable protein, and the living organisms in the soil have proteins in their tissues. Proteins are complex compounds made up of amino acids condensed together through the so-called peptide (—CO.NH—) linkages. These linkages can be broken by chemical hydrolysis to release individual amino acids which can then become complexed with other types of organic matter. Fourteen compounds are commonly found to constitute the bulk of the amino acid content. They are glycine, alanine, valine, leucine, isoleucine, serine, threonine, aspartic acid, glutamic acid, phenylalanine, arginine, histidine, lysine and proline. Many others have been variously reported present, and it is interesting to note that tryptophan does not appear although it is an essential acid in all animal proteins. One or two non-protein amino acids such as γ amino-n-butyric acid and 3.4 dihydroxyphenyl alanine have also been discovered and these are thought to be microbiologically synthesised.

Between about 5 and 10 per cent of the soil nitrogen is associated with amino sugars in which carbohydrate units contain amino groups, as was described earlier, and there is evidence that nucleic acids are present, though in smaller amounts. The remaining nitrogen, something like a half, has not been characterised in any detailed way. It is resistant to extraction by chemical means and deductions concerning its state are based on hypothesis and analogy. Popular theories postulate either the formation of stable complexes between proteins, amino acid, or ammonia with lignin derivatives, or of compounds produced by amino compounds and carbohydrates. All the fixed nitrogen compounds are remarkably stable and in any one season only 2 or 3 per cent is mineralised by biological action whereas freshly incorporated nitrogenous organic matter mineralises quite readily.

The Lignin Fraction

The evidence that lignin constitutes a large proportion of soil humus is largely circumstantial, and direct attempts to isolate and identify lignin by orthodox chemical technique have been unsuccessful although various workers have found a similarity in the ultraviolet absorption spectra of humus and a number of lignins from various sources.

There can be no doubt that the plant lignins suffer significant chemical changes when incorporated in the soil and that oxidative changes are prominent, giving an increased number of carboxyl groups which can in some measure explain the increase in base exchange capacity when lignin is allowed to decompose naturally. Demethylation of the lignin also occurs, possibly due to bacterial action.

In addition to the reactions affecting the peripheral groupings of the lignin it is also possible that condensations and polymerisations involving the cyclic structures and ammonia may take place as has been indicated.

Physical Properties of Humus

Physically humus shows qualities associated with many hydrophyllic colloidal materials. In the presence of basic cations other than the alkali metals, it may assume a flocculated, gel-like form having a high water-holding capacity, capable of expanding and contracting according to its moisture content, although below a critical moisture content it irreversibly assumes a brittle vitreous state. In the absence of polyvalent cations humus may be peptised to give a sol form, assuming a potential mobility within the soil. The colloidal organic matter can associate with the mineral colloids forming, in dilute suspensions, a protected colloidal sol. Even at low moisture contents which give rise to an essentially solid system, the organic and inorganic colloids form intimate associations from which it is not possible by physical means to separate the two components. Organic colloids may also be physically absorbed on to the surfaces of larger soil particles.

The marked hydrophyllic tendencies of soil organic matter are apparent in the organic residues that are incompletely broken down as well as in the more obviously well decomposed amorphous humic material, although it is probably only the latter form which provides the mixed organic–inorganic associations and gives rise to a form of humus often referred to as 'mull' humus. In contradistinction to mull humus is 'mor' humus in which the humification processes have been limited by various factors giving rise to an accumulation of organic matter. This does not readily mix with the inorganic soil

material and usually forms a distinct layer on the surface. All grada-
tions between extreme forms of mull and mor can occur.

Chemical Properties of Humus

A major difficulty in the study of soil organic matter is that of
extracting samples of sufficient purity, free from both soil mineral
matter and also from unhumified residues. That humus is not a
single substance is now obvious and there is the additional difficulty
of separating the complex into recognisable components.

Many techniques have been suggested for separating humus into
its constituent parts but none can be regarded as being completely
successful although each makes a positive contribution to the problem. The first technique, with historic precedent, is the use of dilute
alkali which dissolves appreciable quantities of organic matter, provided basic cations are initially removed. This use of alkali to separate
completely putrified vegetable matter is of long standing; an early
account by Vaquelin in 1797 comments on the dark coloured extract
obtained by treating rotten elm bark with potassium carbonate, and
this and other contemporary observations gave rise to the idea that
humus—a name apparently first used by Théodore de Saussure to
describe 'terreau végétal' or rotted plant material—had the nature
of an acid. Early in the nineteenth century two types of acid were
recognised: acid humus in which the humic acids were in the free
condition as found in peats, and mild humus in which the acids were
rich in bases which had to be neutralised before the humic acids
could be extracted by alkali.

Treatments of suspensions of humus with compounds which
complex heavy metals have been shown to bring into solution a considerable proportion of the organic matter. It can be shown that
reagents such as sodium pyrophosphate, malic acid, citric acid and
E.D.T.A., which efficiently extract polyvalent metals from soil also
extract appreciable quantities of organic matter. Such observations
lead to the conclusion that in the soil these polyvalent cations form
co-ordination compounds with the organic complex which are insoluble. When the co-ordinating cations are removed by a suitable
solvent the organic complex becomes soluble in water and can be
extracted. The use of alkaline pyrophosphates are to be preferred for
this purpose as the inorganic compounds can be removed from the
mixture by simple dialysis leaving the humus extract as an un-
contaminated aqueous solution.

As a result of the many approaches that have been made to examine the characteristics of humus, different types of product are now visualised as being involved. These are given specific names for convenience but it must be realised that they are not single compounds by any means. They are described as acids because they combine with bases and can occur as salts as well as in an acidic form. Four fractions are commonly described.

CRENIC ACID is the water soluble fraction not normally found in any significant amounts.

FULVIC ACID is the fraction of the humus that is soluble in dilute alkali which is not precipitated when the solution is made acid by addition of mineral acid.

HUMIC ACID is the fraction which forms a precipitate when the alkaline extract is acidified with mineral acid. It is often divided into three sub-fractions, namely *hydromelanic* acid which is soluble in alkali, *grey humic acid* and *brown humic acid*. Grey humic acid has a higher nitrogen content than the brown and is supposed to be the agriculturally valuable form of humus being associated with biological activity, and has high absorptive powers. It is composed mainly of well polymerised high molecular weight compounds, well complexed with mineral matter. Brown humic acid on the other hand is composed of non-polymerised lower molecular weight substances, little complexed with mineral matter and relatively inert biologically. Grey humus is always associated with brown although the latter frequently occurs alone. The separation of these two types is based on the fact that the grey is more sensitive to flocculation by electrolytes than the brown.

HUMIN is the partially modified fresh organic matter remaining as the insoluble residue after the soil has been treated with alkali.

A feature of soil organic matter that is of some significance agriculturally is the ratio of carbon to nitrogen in its empirical composition. Fresh organic material of vegetative origin usually has a wide C : N ratio varying according to type from 40 : 1 to 15 : 1. Humification processes which are predominantly oxidative tend to narrow this ratio, carbon being lost largely as respiratory CO_2 from microbiological development, whilst the nitrogen losses are limited by incorporation in microbiological tissue. In a biologically active soil the C : N ratio tends towards figures around 10 : 1, the ratio normally widening as microbiological activity is restricted.

The presence of certain chemically reactive groups in humified organic matter gives rise to further chemical properties of significance. Carboxyl and phenolic hydroxyl radicals give humus acid properties and enable it to form salts with basic ions. The cation combining capacity is considerably greater than that of the inorganic clay minerals, varying between about 300 and 700 milli-equivalents of cation per cent. Due to the partial dissociation of the combined cations they are 'exchangeable' in a manner analogous to the clay exchangeable cations, and if the capacity is fully saturated with basic cations a suspension in water has an alkaline pH and it is acidic when hydrogen saturated. Varying proportions of acid groups combining with bases give a range of possible pH values, and the salt-like complex, being composed of very weak acids and strong bases, forms a strongly buffered system.

Degree of Humification

There is often a noticeable relationship between humus form and the type of soil associated with it. In order to obtain some quantitative correlation attempts have been made to measure the extent to which humification has taken place. One method is based on the claim that unhumified organic matter dissolves if the total organic content is strongly acetylated. This can be done by treatment with acetic-anhydride and sulphuric acid, or by acetyl bromide. By determining the carbon contents of the total organic matter and the residue after acetylation a degree of decomposition number ZG (Zersetzungsgrad) can be derived.

$$ZG = \frac{Ch}{Ct} \times 100$$

where Ch = the percentage carbon in the acetyl bromide insoluble fractions and Ct = percentage total carbon in the organic matter. Using this latter enumeration it has been found possible to differentiate the organic matter of different genetic groups of soil according to the extent of humification, with ZG values of 70 to 80 for chernozems, 45 to 50 for brown earths and 20 to 30 for podsols.

If the ZG value is measured after the soil has been treated with 5 per cent hydrochloric acid to liberate free lignic and uronic acids which are insoluble in acetyl bromide, a new figure ZG^1 is obtained and the difference between the two can be taken as a measure of partially humified organic matter.

Forms of Humus

The general character of humus varies widely. It can range from highly organic peat to mull humus which in many soils is hardly differentiated from the mineral matter. The many forms of humus do not fall into well separated and distinct species but merge from one distinct form to another. Each type has associated with it certain environmental features and characteristics which can be recognised. A number of important types are described below, each of which can be subdivided in detail where critical comparisons are called for.

MULL humus is associated with conditions that induce rapid mineralisation, with the associated formation of organo-mineral complexes between the clay fraction and the products of biological synthesis. It is associated with a mixed microflora, bacteria and actinomycetes predominating, and a large earthworm population. It is usually dark coloured and the structure of the intimate association of organic and mineral is well aggregated and crumbly. Two sub-types are to be recognised. *Calcmull*, which is associated with soft limestone and is almost black in colour, is base saturated and has a C : N (see p. 63) of about 10, and *forest mull*, which occurs in non-calcareous situations and has a brown colour, has a lower base saturation and a slightly higher C : N of up to 15.

MODER humus is an even less base rich form, found on siliceous parent material, which has a moderate mineralisation rate with fungi prominent in the microflora. It does not form intimate complexes with the finer mineral fractions but remains distinct as a well decomposed humus form. It is dark in colour and is weakly cohesive with a pH of about 5 and a C : N of about 20.

MOR humus is less completely modified and tends to be leafy, fibrous and particulate. It is associated with coniferous and heath vegetation and has a high C : N of 30 to 40. It is acid and forms a separate layer on the soil surface. Mineralisation is slow and fungi are dominant in the microflora.

PEAT development takes place under conditions where the rate of mineralisation is very slow and there is a tendency for organic matter to accumulate as the balance between gain and loss becomes positive. Three types of peat can be recognised, two with a predominantly fibrous structure and one with an almost structureless, plastic texture. The fibrous peats are the *oligotropic* which is normally very acid, with a high C : N ratio, formed under conditions of acidity and waterlogging associated with high rainfall, and *mesotropic* which occurs when mineralisation has been halted by waterlogging associated with a high water table. Ground water is often calcareous and then the peat formed is usually base rich with a pH around neutral. The oligotrophic peats are variously known as 'blanket peat', 'raised moss' or 'bog peat' and the mesotropic as Fens, and all contain more than 30 per cent organic matter in the surface layers. *Anmoor* peat is quite different in form to the peats although it is often associated with them. It is formed under seasonally waterlogged conditions where mineralisation is slow and humification pronounced so that the original organic matter loses all form and breaks down into a dark, massive and plastic texture. There is less than 30 per cent organic matter in the surface soil, the typical mineral components being quartz sand grains.

6   Water and soil

Water is perhaps the most important factor influencing plant growth throughout the world. It is also a major factor controlling the genetic development of soil profiles. Soil–water relationships can be such as to lead to the creation of soil or its destruction by erosion. These relationships are complex, influencing physical properties in terms of structure, expansion, contraction and strength, chemical properties in terms of mineral weathering and the supply of available nutrients, and the biological state of the soil in terms of bacterial, fungal and insect populations.

The water content of a soil can vary within wide limits. At one extreme a soil can be completely waterlogged, when all the volume not occupied by solid matter is filled with water, and at the other, after artificial drying, the water content can be reduced to the combined water existing in hydrated mineral species, with the interstitial spaces being completely air filled. Between these limits the pore space can be variously filled with water, and within limits again, that water can move from more saturated to lesser saturated zones.

If water is progressively withdrawn from a fully saturated soil three states of wetness can be recognised. At first water drains freely from the soil under the influence of gravity, and the rate of loss gradually slows down until no more is being lost even though the soil may appear quite moist. During this state the vapour pressure of the water remains essentially equal to that of a free water surface under the same conditions of temperature and pressure. At the stage when no further water drains away the soil is said to be at 'field capacity'. The moisture content at field capacity differs for different soils and even for the same soil may vary slightly as temperature changes occur. Further loss of water can take place by evaporation or by absorption by plant roots and the moisture content decreases still further until a stage is reached when no further loss from these causes can occur. The soil is now at 'wilting point' as plants can no longer obtain the water necessary to meet their needs and they wilt and die from moisture starvation. Growing plants will show signs of moisture stress before this point is reached and wilting may occur but recovery takes place if the water stress is removed; at wilting point no such recovery can take place. Between field capacity and wilting point the water vapour pressure decreases steadily as the moisture content reduces. The moisture content at wilting point, as in the case of field capacity, varies from soil to soil and again can show slightly variations in the same soil as soluble salts in the soil solution vary.

To remove the moisture left below the wilting level artificial means are called for and this residual water has a very low vapour pressure compared with that of the earlier stages. Figure 6.1 diagrammatically illustrates these stages. The importance of Stage 1 centres around the

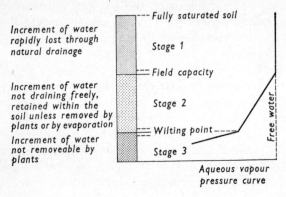

FIG. 6.1   Soil–water stages

denial of access to the soil mass of air necessary for root respiration and the reduction of gaseous diffusion rates which will lead to a build up of carbon dioxide at the expense of oxygen. Stage 2 covers the reserve of available water which becomes depleted as a result of evaporation and depletion by plants. The size of this available water reserve is important and varies from soil to soil and it is within this range that problems of movement of moisture from moist to dryer zones in order to keep transpiring roots supplied are important. The final Stage 3 is somewhat academic in character but its interest lies in its possible illumination of the properties of the particle surfaces with which this water is intimately connected.

These qualitative states of soil water are not defined in terms of moisture contents, as they vary from soil to soil, but need some kind of quantitative interpretation if any real meaning can be attached to them. The behaviour and properties of the water in soil are affected by the size of the particles of which the soil is made, by the state of packing, by the nature of the particles themselves, by the exchangeable ions and by the presence or absence of soluble salts.

The solid phase of the soil mass influences soil water in two ways.

The inorganic and organic colloids of the clay fraction are strongly hydrophyllic and have a high affinity for water molecules and, because they are negatively charged systems, they adsorb the molecules initially as an orientated surface layer. This is because water molecules have a structure in which the electron and proton charges are at opposite sides, giving a polar molecule. Hence when adsorbed on to a negatively charged surface they align themselves with the positive side of the molecule adjacent to the adsorbing surface, presenting a further negatively charged exterior which in turn attracts a further layer of water and so on. Each additional layer of molecules adsorbed is less formally orientated until at about 8 or 10 layers from the surface they become more or less randomly arranged. Because these initially adsorbed layers are under the influence of the electrical field of the colloid they lose their free fluid properties and are quite difficult to remove. As the water films become thicker the fluid properties become normal and moisture movement becomes feasible.

In a solid mass of particles water can only occupy the fraction of the total volume not occupied by solids, that is in the so called 'pore space' of the soil. This pore space is an interconnected system of spaces between the soil particles which because they vary in size and shape give rise to a complex network of pores equally variable in shape and size. In very coarse sand, the system will be one of spaces linked together through narrow necks where the particles touch and in clay soils it will similarly have spaces between the clay particles linked together by smaller passages, although in the latter case all the pore spaces will be on a much smaller scale. In mixed loams or where clay soils are made up of aggregates a much wider range of pore sizes will be present.

As soil minerals are all strongly hydrophyllic the contact angle of liquid water and mineral particle will be zero, and as the soil mass contains this continuum of pore space with voids of irregular shape and cross section, water will be drawn into the soil mass in the manner of a capillary tube where the relationship

$$h = \frac{2\eta \cos \theta}{gr\varrho}$$

holds where h represents the vertical height to which water rise in the capillary tube of radius r, η is the surface tension of water, θ is the angle of contact, and ϱ the density of water and g the gravitational constant.

As the angle of contact in this case is zero, the expression can be rearranged to read

$$hg\varrho = \frac{2\eta}{r}$$

where $hg\varrho$ represents a force; and as 2η can be taken as constant we find that the force with which water is drawn into a capillary tube is inversely proportional to the radius of the tube.

If soil is regarded as a simple physical system containing a mass of pores in contact with water it would be expected to draw water into itself on this capillary attraction basis. This can easily be demonstrated to be true by standing the ends of glass tubes filled with dry soil in a reservoir of water when the water will rise up the tubes. The rates at which the water rises and the heights reached vary very much and whilst in theory the rise in a clay soil should be much greater than that in a sandy soil it takes place much more slowly due to closures of pores by the swelling of clay minerals so that a simple comparison is often hard to achieve.

Water can also be withdrawn from a moist soil by applying a suction force in some appropriate way. If a soil is very wet small suction will extract water, if it is dry much suction may be needed to remove even an infinitely small portion of the water present. This means that wetness of soil can be assessed by measuring the suction needed to extract water from it. This concept recognizes an energy relationship between water and soil, the force required to move water varying with moisture content. The force required to abstract water from a soil at a given moisture content is termed the 'moisture potential' and this total force is the sum of a capillary potential and an osmotic potential. The capillary potential can be regarded as the reverse of the capillary rise effect described above, the force being inversely proportional to the capillary tube radius. Thus in a system of mixed pore sizes fully saturated with water, a given suction will empty the larger pores first, leaving the finer ones still filled. As the suction gradually increases the pores will progressively empty until only the finest remain filled with water and at extremes of suction even these will become empty.

The osmotic potential relates to solutes which may exist in solution within the soil moisture. These will have an osmotic pressure which will act to draw pure water into the soil solution. This effect will be additional to the capillary attraction but will only be of significance when soil moisture contents become very slow and in most soils can often be neglected.

At higher moisture contents it is possible to measure moisture potentials by direct methods involving the application of suction to soil, relating the suction force to moisture content. The technique can well be appreciated by considering a water manometer (Fig. 6.2), one arm of which is open to the atmosphere, the other arm being attached to a container in which a layer of soil is supported on a porous plate. The manometer is completely filled with water, saturating the soil. If the free arm of the manometer is lowered, the soil will be subjected to a negative hydrostatic pressure, or suction, and

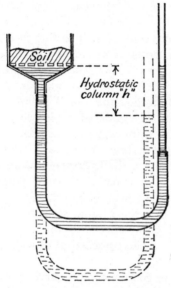

FIG. 6.2 Manometric apparatus for suction pressure measurement.

water will be drawn from the soil, through the porous plate, to neutralise this negative pressure. Provided that the plate is sufficiently finely pored for the surface tension of the water films in these pores to withstand the suctions applied without the films breaking and letting air enter the manometer tube, an equilibrium will be established at the soil/plate interface between the suction force applied and the force of attraction of the soil for its water. The moisture contents of soil at different levels of suction force (moisture potential) can thus be measured and the relationship graphically presented. This technique is somewhat limited in its application as it is not practicable to apply suctions approaching or exceeding atmospheric pressure. For higher capillary potentials it is necessary to use either a direct pressure technique, using compressed air to push out soil moisture, or centrifugal methods in which the gravitational force can be increased many thousand times. Indirect methods involving vapour pressure and freezing-point determinations must be used to estimate extremely high capillary potentials.

It follows from what has been said that the lower the water content of a soil the more strongly is it held and the greater its moisture potential, and in a soil heterogeneously wetted water will always move from places of lower moisture potential to places of higher potential although the speed of such movement can be very slow in many instances.

In relating moisture content with suction force, it is customary to define the force in terms of a hydrostatic column rather than in absolute units; thus if a soil establishes a moisture equilibrium with a hydrostatic column 1,000 cm long (roughly equivalent to a pressure of one atmosphere) its moisture would be said to have a capillary potential of 1,000 cm. The graph of suction against moisture content is often referred to as the moisture characteristic curve.

Schofield suggested a more convenient notation in which the logarithm of the height of the hydrostatic column defines the force with which moisture is retained in the soil. This function he designated pF. The use of pF has the advantage of reducing the linear range of pressures to a logarithmic scale convenient for graphical representation, and does away with the implication in the older term that capillarity is the principal phenomenon involved in soil–water relationships.

Many workers have constructed pF curves and found that there is a hysteresis effect when drying and wetting curves are compared for the same soil; thus a soil brought to a certain pF by drying a wet soil will have a higher moisture content than if brought to the same pF by wetting a dry soil. No doubt the two percentages would be the same if an indefinitely long time were allowed for the very slow attainment of a perfect equilibrium, but for practical purposes there is a drying moisture characteristic and a wetting characteristic. Fig. 6.3 shows pF curves obtained by Schofield for a Rothamsted soil.

The use of pF does not in itself imply any mechanism whereby the soil attracts moisture over the range of possible moisture contents. It merely gives a reflection of a continuous energy relationship between the soil and its water, and can be used to define moisture

status in numerical terms indicating the ability of soils to hold water
and to supply water to the plant.

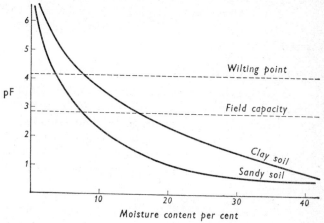

Fig. 6.3   pF curves.

The Moisture Constants of Soil

The determination of a full moisture characteristic is a long and
difficult procedure and in order to obtain some idea of the moisture
status of a soil there are a number of so-called moisture constants
which can be determined relatively easily and may be used to help
characterise soil/moisture relationships. The more important of these
which take historic precedent over moisture characteristics are
described below.

The Maximum Water Capacity

This is the most obvious and most easily measured. It is the maximum amount of water which the soil can hold when drainage is
entirely precluded and the air is wholly displaced. It is, in fact, the
pore space (see p. 69).

The Moisture-holding Capacity or Field Capacity

The maximum amount of water which a soil can hold when it is
drained as freely as its texture will allow is the moisture holding capacity. It is the maximum amount of water which can remain in a soil
when all external factors which deter drainage have been removed,
and when the only hindrance to free drainage is the soil itself.

The actual moisture contents at field capacity vary considerably from soil to soil. There is also a variation with depth and with the height of the soil above the water table. For sandy soils values of about 20 to 25 per cent would be typical, for clay loams 25 to 35 per cent, and for organic soils the moisture content at field capacity can easily exceed 50 per cent.

The Hygroscopic Coefficient

When a dry soil is in contact with an atmosphere saturated with water vapour, the percentage of water taken up by the soil is the hygroscopic coefficient at that temperature. The amounts of water absorbed under such conditions range from about 2 per cent in light soils to 13 per cent or more in heavy soils.

Two suggestions have been made about the value of the hygroscopic coefficient:

(1) It has been suggested that the water in excess of this value is the water available for the plant, and thus if the hygroscopic coefficient is subtracted from the total water content of the soil, a measure of available water is obtained.

(2) The value of the hygroscopic coefficient as a measure of the surface of soil particles has been discussed, originally by Mitscherlich, who held that it evaluates the soil texture in a better way than mechanical analysis.

There are, however, many difficulties in the way of evaluating the worth of the hygroscopic coefficient. Besides the practical difficulties of the actual determinations, there is the extreme difficulty of interpreting results arising from the complexity of the soil surface and the impossibility of isolating a pure hygroscopic effect from all secondary actions.

Some determinations have been made of the heat developed when a dry soil is wetted, and attempts have been made to correlate heat of wetting and hygroscopic coefficient. The same difficulty arises, namely that the production of heat during the wetting of soil is a complex of chemical and physical actions.

The Moisture Equivalent

Another experimental method which has been adopted in the search for characteristic soil moisture constants is the removal of as much as possible of the water by some arbitrarily fixed and measurable force, and the estimation of the remainder. Introduced by Briggs and McLane, this constant is the percentage of water remaining after centrifuging saturated soils, in perforated cups, at a speed such that a centrifugal force equivalent to 1,000 times the force of gravity is applied to the mass for a given time, usually 30 minutes. Many empirical relationships have been evolved featuring moisture equivalent, including its correlation with mechanical analysis and its equation with field capacity. Some such latter relationship, if universally reliable, could be most useful in giving a laboratory method for assessing field capacity and thereby facilitating the estimation of a soil's potential for retaining water for crop utilisation. However, the empirical nature of these uses of moisture equivalent makes generalisations unwise, and specific associations to deal with particular cases are necessary.

The Wilting Coefficient

None of the determinations so far considered directly involves the plant. The measurement of a soil moisture constant which directly involves a plant effect was attempted by Briggs and Shantz, who determined the amount of water still remaining in soils when 'permanent' wilting of plants grown in them had just set in. These workers endeavoured to correlate wilting point with all other moisture constants, but experience again shows that no hard-and-fast relationships exist. It is, however, agreed that it is a valuable datum point in considering soil moisture studies, although, as with all the other so-called moisture constants, it is only one point on the curve describing the energy state of soil moisture as its proportion relative to the solid phase changes.

The Bar

The moisture constants are useful single figure values to describe any particular soil and to help predict its moisture status under field conditions. In the general context of soil work it is more suitable to use some equipotential value so that all soils can be compared on the same footing. The pF notation described earlier is useful in this context. Modern investigators frequently adopt particular suction pressures to designate stages of wetness and use the 'bar' as the fundamental unit. A bar is a unit of pressure equal to one million dynes per square centimetre, which is approximately equivalent to a pressure of one atmosphere. Thus, the wet limit for water available to plants under natural conditions (the field capacity) is taken as the 'one third bar percentage' which is defined as the percentage of water contained in a soil that has been saturated, subjected to, and is in equilibrium with an applied pressure of $\frac{1}{3}$ bar. It corresponds to a pF value of 2·53. In similar terms the wilting point is equated with the 'fifteen bar percentage' which is equivalent to a pF value of 4·18. The hygroscopic coefficient corresponds to about thirty bar pressure or pF 4·5.

THE MOVEMENT OF WATER IN SOILS

Within the soil, water moves through those spaces not occupied by solid particles, in other words through the pore spaces. If the soil is composed largely of single coarse grains, the individual pores will be large, whereas in fine grained material the pores will be small. In

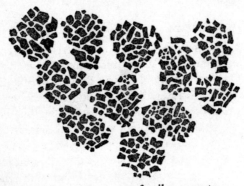

FIG. 6.4   Diagrammatic arrangement of soil aggregates.

the latter case the pore sizes can be increased if the small individual particles are grouped together to form aggregates or crumbs, as each aggregate will act as a coarse particle to give large pores between aggregates, while at the same time there will be small pores within the aggregates (Fig. 6.4).

It must be remembered that pores within a soil are not isolated pockets, but form a continuous system of channels radiating in all directions, and that these channels have very variable cross-sectional areas.

If a saturated soil with an accumulation of water on its surface is considered, water will flow downwards under the influence of gravity. The rate of flow through the pore channels will decrease as the diameter of the pore channels decreases, since frictional forces tending to retard the flow will become increasingly prominent. So, in the finest of the pore channels the rate of gravitational flow will become infinitesimally small. This point emphasises the necessity for developing a crumb or aggregate structure in heavy clay soils in order to facilitate drainage. This picture envisages a continuous water column flowing through the soil, different parts of the column moving at different speeds.

Differences in the ease with which water flows through pores of different sizes is often acknowledged by classifying them into two

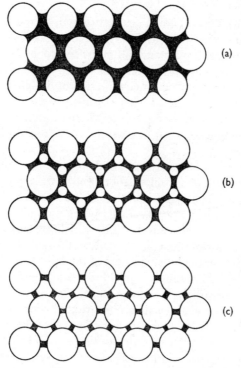

(a)

(b)

(c)

FIG. 6.5 Stages in particle water systems. (a) capillary, (b) funicular, (c) pendular.

groups referred to as 'capillary' and 'non-capillary pores'. The former include the smaller range of pores which remain water filled when free drainage has brought the soil to field capacity, and the latter those which are too large to retain water under similar conditions. The equivalent mean diameter separating these groups is 0·05 mm.

As the total moisture decreases the water in capillaries will retract until a continuous air column is introduced, although, at the same time, there will exist a continuous moisture film around the soil particles. Using the nomenclature of Keen, the moisture condition has now passed from the 'capillary' stage to the 'funicular' stage. On further reduction in moisture content the funicular stage passes to the 'pendular' condition and the continuous film of liquid water becomes broken, persisting only as annular rings around the points of contact of the soil particles. Fig. 6.5 gives a simplified picture of these conditions.

In the capillary stage, i.e. in a saturated soil, the rate of movement will be limited by the width of the pores and the viscosity of water; in the funicular stage, movement will occur to reduce any inequalities in the energy states of water molecules within any one environment, which can be interpreted as an equalisation of like elements of water film throughout the system. This being so, water will move from regions where the films are thickest to where they are thinnest. In the pendular stage movement of free water is not possible as there is no continuity of path.

This appreciation of water status and movement enables an understanding of the fate of a limited amount of water applied to the surface of a dry soil. On saturation of the surface layers drainage will take place through the larger pores, smaller pores remaining filled because of the surface tension of the water. The initial drainage will be drawn into the finer pores immediately below the surface by capillary forces, and will wet the inner surfaces of the coarser pores until there is no longer a sufficient supply available for bridging the gaps between the annular rings of the pendular stage. Downward movement of water then ceases and the water in the wetted zone begins to redistribute itself to unify the moisture films, the very thin films attracting moisture from the finer capillaries until an equilibrium is reached. At equilibrium the soil will be at field capacity, and the depth of the wetted zone will be related to the amount of water available, and below the wetted zone the soil will remain quite dry.

Upward movement of water can take place from a zone of saturation such as a water table by capillary rise, but such rise is considerably limited by frictional considerations, by the fact that capillaries are variable and irregular in diameter, and by closure of pores due to the swelling of soil colloidal material on wetting and rarely is there found a rise of more than a metre due to capillarity.

Lateral movement of water in soils is brought about by the forces tending to equalise the energy status of water throughout any particular zone, and takes place very slowly largely on account of the physical impedance which is common to water movement generally within soil.

These considerations of water movement have been on the movement in the liquid phase. Superimposed upon this is also movement of water in the vapour phase. Movement of water is brought about when there exist differences in vapour pressure, water vapour moving from regions of high vapour pressure to regions of low vapour pressure. It is in effect a distillation. Within a soil, other things being equal, water vapour will diffuse from regions of wetness to regions of dryness, assuming that there is free passage for the vapour. The importance of this effect is much disputed and little quantitative evidence is available to assess its significance. It might be expected to vary from soil to soil according to texture, and, by virtue of the view that soil air is normally saturated with respect to water vapour, deficiency will only be found in the driest soils. The influence of temperature on vapour pressure and on the capacity of air to contain water vapour may be of greater import. Air saturated with water vapour at 20°C holds 17·1 g of water per cubic metre; at 10°C it holds 9·3 g of water and at 0°C 4·8 g. If the soil air temperature is reduced water will be condensed from the vapour phase, being converted to the liquid state. The existence of a temperature gradient will also cause water vapour to move from the warmer regions of high vapour pressure to the cooler regions where the vapour pressure is lower, where it will condense. This means that in winter when the surface of the soil is cooler than the soil at depth, there will be an upward movement of water via the vapour phase and in late summer a reversal of this effect.

7   Air and soil

It is normal for soil to contain, in addition to the mineral and organic constituents, air which, together with the water as described in the previous chapter, occupies the pore space of the soil. This air plays a not inconsiderable part in determining the root development of growing plants and is on that account alone an important soil constituent in its own right.

Amount of Soil Air

The volume of air contained in a soil will depend upon the soil's porosity and its moisture content. The total pore space normally ranges from about 30 per cent to 50 per cent by volume, heavier soils usually having the higher values. This value decreases with compaction and hence with soil depth. The pore space too is variable with respect to its configuration and is frequently divided into two categories, capillary porosity and non-capillary porosity, the former comprising the smaller pores in which water is tenaciously held by surface tension forces, the latter including the larger sized pores from which water drains reasonably well.

The total pore space in soil can be measured indirectly by determining the true and apparent densities of a block of soil. If ϱ is the true density of the soil solids and d the 'apparent' density or density of the whole block of soil, then for 100 g of whole soil the volume will be $\dfrac{100}{d}$ and the volume of solid particles contained therein will be $\dfrac{100}{\varrho}$. The volume of 'voids' will therefore be $\dfrac{100}{d} - \dfrac{100}{\varrho}$ and the volume percentage of pore space becomes $100\,\dfrac{\varrho - d}{\varrho}$. It is also possible to measure pore space by invoking Boyle's law and relating changes in gas pressure to changes in volume occurring when a known initial volume of soil is compressed.

The air in field soils will usually occupy the moisture-free spaces in the non-capillary ranges of the pore space, extending, as indicated in Chapter 6, to the capillary pores as the moisture content decreases.

The interrelationship of air, water and pore space indicates that the air content of a soil will decrease with depth as the proportions of capillary pores and the water content increase, and at the water

table it will become zero. In the surface soil the air content w
greatest in light soils and in soils with a good crumb structure, s⸏‿
soils having a high non-capillary porosity.

PORE SIZE DISTRIBUTION

Soil pores form a continuum of varying dimensions with large and
small pores interspersed and interconnected, and it is useful to have
some idea of pore size distribution in soils. It is often sufficient to
differentiate between two ranges of pore size, referred to as capillary
and non-capillary pores as mentioned above, but occasionally a
continuous pore size distribution curve is required for a fuller
interpretation of soil structure phenomena.

Using an apparatus similar to that described for determining suction pressure (see page 71) it is possible to measure the water withdrawn from a saturated soil at any particular suction pressure, and
this will equal the volume of pores having a greater radius than that
corresponding to the suction pressure (page 69). By making a series
of such measurements it is possible to draw a graph relating pore
size to moisture tension and the slope of the curve at its various points
reflects the pore size distribution. When this is done for normal arable
soils two pore size peaks are often found reflecting inter-aggregate
and intra-aggregate pore sizes.

AERATION POROSITY

In order to obtain a basis for comparing the relative aeration capacities of soils, a quantitative standard has been suggested and termed
the *aeration porosity*. It is defined as being the proportion of the total
soil volume occupied by air when a moist soil is in equilibrium with a
suction pressure equivalent to a 50 cm column of water. It corresponds to a stage of aeration when pores greater than 0·06 mm in
diameter will be filled with air. Soils having aeration porosity values
of 10 per cent or more are usually quite adequately aerated to
meet arable crop requirements. It should be noted that in natural
soils pores of quite large size often occur which are completely
surrounded by much smaller pores. These may remain filled with air
when the soil is submerged or remain filled with water when the soil
is drying out when subjected to tensions which would normally
empty them, because of restrictions imposed by the small pores.
Such pores are often referred to as 'blocked' pores.

PORE SIZE DISTRIBUTION AND AERATION POROSITY

Composition of Soil Air

Compared with the free air of the atmosphere, the average composition is not greatly different, as will be seen in the following figures comparing the dry air compositions. The figures are percentages by volume.

	Soil air	Atmospheric air
Oxygen	20·3	20·99
Nitrogen, etc.	79·2	78·95
Carbon dioxide	0·5	0·03

There are, however, three points to be mentioned in connection with the comparison.

(1) Soil air is generally saturated or nearly saturated with water vapour.
(2) The average carbon-dioxide content of soil air is many times as great as that found in free air.
(3) There is a considerable variation in the composition of soil air, both within the soil itself, and from time to time; much more than is found to occur in the atmosphere.

Carbon-dioxide Content

The main cause of the comparatively large percentage of carbon dioxide in soil air is the activity of micro-organisms. It is possible to demonstrate this in two ways.

Firstly the treatment of the soil with antiseptics prohibits the further production of carbon dioxide.

Secondly, the fluctuations in the amounts of carbon dioxide are directly correlated with the fluctuations in the amounts of other products which are known to be the result of micro-organic activity and follows the seasonal fluctuations associated with such activity. Several workers have demonstrated a relationship between the production of carbon dioxide and of ammonia for instance, and a direct correlation can be shown to exist between the bacterial numbers occurring in the soil, the amount of nitrate produced and the carbon-dioxide content. In addition to the bacterially produced carbon dioxide, a small amount is contributed by root respiration which is also taking place during the growth of these plant organs. Respiration also involves oxygen consumption so that within the soil all respiring organisms will be using oxygen and the oxygen level of the air will tend to be reduced. It is not easy to cite precise figures for the production rate of carbon dioxide as it will vary according to the prevailing micro-organic activity which will be dependent upon temperature, moisture status, organic matter, etc. Under natural conditions, however, figures of the order of 10 litres of carbon dioxide per day per square metre of soil to a depth of 20 cm might be expected. Oxygen consumption corresponding to this would be in general of the same order.

The soil air is in contact with the atmospheric air and there must therefore be a continuous diffusion tending to equalise the composition of the whole connected system. In certain exceptional circumstances this diffusion may become pronounced and dominate the production of carbon dioxide. Generally, however, the production of carbon dioxide which accompanies the activities of micro-organisms is the dominant process, and consequently when diffusion and production are in equilibrium there is a higher content of carbon dioxide in the soil air than in the atmospheric air.

Provided that the pore space is freely open to the atmosphere the higher concentration of carbon dioxide in the soil air will tend to equate itself with that in the atmosphere by straightforward gaseous diffusion in accord with Graham's law of diffusion of gases. This states that the velocity of diffusion is inversely proportional to the square root of the density of the gas, a law which also holds for rate of passage of a gas through fine openings. The densities of carbon dioxide, oxygen and nitrogen at N.T.P. are 1·9769, 1·4290, and 1·2506 grammes per litre respectively, so that for similar inequalities of partial pressures of these gases the carbon dioxide will have the slowest diffusion speed. Simple application of diffusion laws is, however, not very easy, as the system is a complicated one with a variable generation rate of carbon dioxide and where changes in temperatures of the soil and in barometric pressures of the outer atmosphere may cause expansion or contraction of gases influencing gaseous transfer between the soil and the air above.

It follows that the more easily gaseous diffusion occurs the lower will be the content of carbon dioxide in the soil air, and conditions that make diffusion more difficult will cause the carbon-dioxide content to increase. Diffusion effects will be more marked near the surface of the soil and in general the carbon-dioxide content will increase with soil depth as the pore spaces and channels are more restricted at depth and more likely to be partly filled with liquid water.

SOIL AIR AND SOIL FERTILITY

The soil air must supply adequate oxygen for plant root respiration and to meet the needs of the soil micro-organisms. As the air is distributed through the interconnected network of soil pores linked to the atmosphere gaseous diffusion between the two systems will take place so long as the link is maintained. As the moisture content of the soil increases, the ability of the soil to meet the oxygen requirements is reduced and soil fertility can be considerably reduced also.

In the absence of oxygen, or when it is present in minimal amounts a number of possible consequences may happen. They relate mainly to microbial activity and can involve the reduction of nitrate to nitrite and hence to gaseous nitrogen with a lowering of the nutrient level in the soil or the production of a number of metabolic by-products not normally found in aerobic environments. These include butyric acid, sulphides, ferrous iron, methane and ethylene all of which prevent root growth even when the amounts present are trace quantities.

The direct effects of low soil oxygen have perhaps been over-exaggerated in the past. Transport of oxygen through continuous gas channels within the plant that extend from leaves to roots can occur, and root respiration requirements can adequately be met by this mechanism as instanced by the survival of bog plants and rice.

However, as the soil oxygen content decreases, the carbon dioxide content increases and the possibility of carbon dioxide toxicity occurs. Modern evidence suggests that this is unlikely to occur until the partial pressure of the CO_2 in the soil air exceeds that of O_2, and this critical value will be about 0.1 atmosphere.

Within a continuous network of air filled pores gaseous diffusion is sufficiently rapid in most instances to maintain adequate aeration. Diffusion through water filled pores is much slower by a factor of 10,000 times. Where, therefore, water filled pockets occur within the soil, as perhaps within aggregates of a moist soil, oxygen free zones may possibly develop leading to localised production of root toxic substances.

This has been shown to be unlikely in natural soils until the gas filled pore space falls to a value below 12 per cent unless large water-filled aggregates exist in the soil. Calculations matching measured respiration rates under average British arable conditions to the diffusion rates of oxygen and carbon dioxide in water filled pores suggest that these water-filled aggregates would require to be greater than two centimetres in diameter at a total gas space of 12 per cent before the possibility that poor aeration could significantly restrict plant growth.

However, where diffusion is limited by the fact that the pore spaces are water filled then the introduction of drainage facilities will help; the cultivation of a crumb structure to heavy textured soils will increase the size of pore openings and channels, making gaseous interchange easier. These are inherent improvements, but *ad hoc* conditions sometimes occur to inhibit diffusion and cause a build-up of carbon dioxide concentration. The 'capping' of the soil surface after heavy rain forms a crust of impermeable soil through which gases diffuse only slowly. Light rain falling on a dry soil also reduces the permeability of the upper soil layers by filling the capillary pores and forming a water seal.

Rain-water falling on the soil does however assist in some small measure to regenerate the oxygen content of the soil air as it contains a significant quantity of this gas in solution. The quantity is fairly small, contributing in this country about 10 litres per square metre of surface per year. The biological requirements of soils for oxygen on the same basis may amount to anything between 1,000 and 15,000 litres per year.

8   The soil profile

The processes involved in the formation of soil parent material from the solid rocks of the earth have been discussed in Chapter 2 when the weathered products were shown to be the resultant of a number of possible processes acting in concert. These processes can be classed together and described as geochemical weathering processes.

The further natural evolution of the parent material to form soil is the result of the superposition of additional development factors on the chemical and physical decompositions already described, and the whole can be referred to as pedochemical weathering. The introduction of these new factors, which are initially of a biological character, does not necessarily impede the geochemical weathering which is, within limits, more likely to be accelerated as soil development proceeds.

When a vertical section of a well developed soil is examined it exhibits a differentiation into distinctive horizontal bands which often differ both in appearance and in chemical constitution. Each of such bands is known as a soil horizon and the section from the surface to the unchanged parent material beneath is known as the soil profile or solum. In the pedological sense soil development and soil profile development are to be regarded as synonymous, the lower horizons being as much a part of the soil as the surface layers. This differentiation into horizons is a fundamental feature of all soil profiles wherever they may occur, and the horizon characteristics of different kinds of soils enable these to be placed into type categories which will be discussed later.

The processes involved in horizon development are well illustrated by examining the case of soils developing under a humid temperate climate, and may be considered under three main headings: accumulation of organic matter in the surface layers; leaching of the profile (eluviation); and the deposition of leached constituents (illuviation).

Organic Matter Accumulation

The introduction of biological influences into weathered mineral residues can be said to initiate soil formation. The nature and rate of biological modification vary according to a number of environmental factors, including the character of the parent material, whether it be rich or poor in basic cations, or whether it be a coarse free-draining material or finely divided and compacted. Topographical combined with climatological factors give rise to differing temperature and moisture characteristics, all of which influence the development of the natural flora and fauna.

In the initial stages, the introduction and establishment of an elementary biological population has the effect of stabilising the weathering surface, and this early population may be of many kinds. Under montane conditions, for example, where the parent material is only slightly altered rock detritus, mosses and lichens are the prominent colonising genera, followed by an enrichment of the flora after a stable root anchorage has been established to include ferns, grasses and flowering plants. The precise nature of the more varied flora

TABLE 10

	Bacteria per g oven-dried material	Fungi per g oven-dried material
Open sand	18,000	270
Yellow dunes (Ammophila arenarea)	1,630,000	1,700
Early fixed dunes (A. arenaria with other grasses)	1,700,000	69,470
Dune pasture	2,230,000	109,780

varies according to the inherent ability of the mineral residues to supply the nutritional need of the potential colonists and the degree of exposure to which they are submitted. On transported soil parent material, as for example during the development of sand dunes, the original colonising species is often marram grass, forming initially open colonies which spread and thus eventually stabilise the surface. This permits the establishment of other grasses such as creeping fescue and various pasture types.

At the same time, as colonisation proceeds, an increase in the microbiological flora takes place as illustrated in a particular example of this type by the investigations of D. M. Webley *et al.* on a dune site survey in Scotland. Table 10 gives figures from their work illustrating this point.

As succeeding generations of the colonising species develop and die organic matter is introduced into the surface and into the surface layers, supplemented by the increasing development of microorganisms and their residual products. The organic debris may accumulate on the surface, giving rise to the uppermost defined soil horizon which is known as the O horizon, although in many soils this horizon may be absent due to the fact that the rate of decomposition of the surface litter is sufficiently high to prevent its accumulation.

Differentiation within the O horizon is often possible and frequently at least two divisions can be recognised although they merge imperceptibly one into the other. Uppermost is the scarcely altered residues in which the morphological characters of the original sources may be easily recognised. This is called the L layer or litter layer and it merges into the F or fermentation layer in which partially modified litter becomes noticeably present. Underneath this the decomposition processes associated with humification come into play, and the character of the organic remains changes, with loss of physical structure, the whole becoming homogeneous to give a humus-like material which constitutes the H or humus layer.

The factors controlling the development of the O horizon, apart from the physical forces of erosion by wind or water, are those factors which influence humification. Conditions under which microbial action is depressed will tend to favour the accumulation of surface litter. Where a high water table persists, giving rise to anaerobic conditions inimical to bacterial action, the formation of an O horizon will be encouraged, and if the height of the water is such as to give rise to surface water the accumulation of organic debris may cause the formation of peat beds. If the mineral material beneath has low base status and conditions of acidity develop, contributed to by the organic acids formed from the decaying litter, again a suppression of bacteria will ensue with the possible accumulation of a surface matt. The soil fauna also plays a significant part in the establishment of an O horizon. Under neutral and slightly alkaline conditions, earthworms, snails, millipedes, etc., establish themselves and play an important role in modifying and mixing the surface litter with the lower horizons of the soil, whereas under acid conditions such species are inhibited and mites, diptera larvae and beetles typify the fauna, which does not significantly change the character of the horizon.

Temperature influences are often important in the development of O horizons, high temperatures favouring rapid decomposition of organic matter and low temperatures slowing down humification rates with a consequent accumulation of surface residues.

The nature of the residues themselves plays a significant part in this connection. Highly lignified woody residues and those containing significant contents of waxes and tannins subdue bacterial activities and decompose only slowly, tending to accumulate. Conifer needles, heather, and ling are characteristic of those types, whilst deciduous litters, in which there is normally a high content of alkali and alkaline earth cations to check excess development of acidity, decompose more quickly to give *mull* humus. Moreover, as this type of vegetation is usually associated with base rich soils and their associated fauna, rapid incorporation of the humified material with the soil usually occurs.

This gives a soil horizon of intimately mixed organic and inorganic matter (normally predominantly inorganic) which is described as an A horizon. It is the horizon of maximum microbiological activity and is usually dark coloured as a result of the humified organic fraction. If such a horizon is found under arable cultivation it is described as an Ap layer to indicate disturbance by ploughing.

The type of semi-humified organic matter that normally constitutes the O layer is often Mor humus and is characteristically acid in character although the term Mor was originally confined to forest litters of this kind as found in coniferous forests.

Eluviation

The downward movement of water through the soil parent material, carrying with it the soluble decomposition products of the O horizon, is responsible for a further stage in the development of the soil profile. Figuring largely in the decomposition products of organic matter are acids which, together with dissolved respiratory carbon dioxide of micro-organisms, give an acid pH to the percolating waters. This increases their solvent action and catalyses the hydrolysis of primary minerals which occurs in the normal weathering process. The continued action of such percolating water has a leaching effect on the upper layers of the soil profile, removing soluble material and modifying the residue to form the E horizons or horizons of eluviation.

The intensity of the leaching effect is most obviously reflected in the cation status of the soil mineral material. The hydrolytic weathering of many primary minerals is accompanied by the liberation of alkali and alkaline earth cations and the formation of secondary minerals, including the clay minerals (Chapter 2, p. 16). These clay minerals retain a proportion of the liberated cations in an exchangeable form, whilst under conditions of free drainage the surplus is leached from the system and eventually finds its way to the sea. In the initial stages of leaching, therefore, free alkali is removed from the profile although the exchange capacity of the clay complex will be filled by basic cations. If calcium carbonate is present as a mineral constituent of the profile, this will tend to dissolve in the slightly acid water. The calcium ions will compete for the exchange places of the soil colloids, and, because of their high concentration relative to the alkali cations, will tend to dominate the exchange complex. The predominantly calcium-saturated soil will normally persist so long as there is present free calcium carbonate. When the free calcium carbonate is completely dissolved, further percolation by the acid leachate will lead to a modification of the exchangeable cations of the clay, hydrogen tending to replace the basic ions in the order Na, K, Ca, and Mg, although no one exchange reaction proceeds to the exclusion of others. The leached cations, accompanied mainly by bicarbonate and nitrate anions, are then washed out of the E horizon. These exchanges lead to a general lowering of the profile pH, and under acid conditions the hydrated oxides and hydroxides of iron may dissociate to give iron ions and as the acidity becomes fairly intense clay lattices suffer partial breakdown with the liberation of aluminium ions: these ions may then assume the role of exchangeable cations and achieve mobility.

The effect of the leaching process can therefore be summarised in a number of consecutive stages—not of course clear cut, but sufficiently distinct to be separately enumerated:

(a) loss of soluble alkali salts if present,
(b) solution of free calcium carbonate and near saturation of exchange complex with calcium ions,
(c) loss from the exchange complex of alkali and alkaline earth cations,
(d) mobilisation of iron and aluminium.

Horizons within the soil profile which have been subjected to such losses of constituents are referred to as E horizons.

Restrictions on the leaching action can occur anywhere within the sequence of stages for reasons such as drainage impedance or recirculation of cations by vegetative growth, and the rate of development will be dependent on other factors such as climatic variations and the physical properties of the profile material affecting the percolation process.

In addition to a loss of cations from the E horizons during leaching, their organic matter status may also be affected, and under conditions of intense leaching organic matter may be almost entirely washed from the upper horizons of the profile. Such movement of humus is normally associated with movement of iron and aluminium sesquioxides and occurs under conditions of high acidity. The colloidal organic matter which, in the presence of basic cations, is normally coagulated, becomes peptised and passes into a sol form under which conditions it can move down the profile. The associated movement of sesquioxides may be due to a variety of causes. Physical peptisation of sesquioxides under the influence of humus with the formation of 'protected' sols is thought to occur, as also is the rendering soluble of the iron and aluminium oxides by simple polybasic and hydroxy acids from the decomposing organic matter. The clay content of highly leached E horizons is usually low, due to peptisation of the acid clays and their mechanical washing down the profile.

It is often possible to recognise different types of E horizon according to which particular constituent has been predominantly leached away. Where this occurs the different E horizons have a qualifying suffix letter added to the symbol E. Thus Ea horizons have lost sesquioxides as a result of leaching, whilst Eb horizons have had some of their original clay content washed out.

Illuviation

The products formed during the leaching of the surface layers are not necessarily lost to the profile as a whole, but pass down to regions where the rate of geochemical weathering is low and where the biological influences are at a very reduced level. As the environment changes the mobility of the translocated constituents of the E horizons may be checked or stopped altogether when accumulation of constituents occurs. Horizons formed as a result of this process are known as B horizons. According to the degree and type of eluviation the enrichments may be calcareous or ferruginous or they may consist of organic matter or sesquioxides or simply enrichments of the clay fraction by mechanical impedance of elutriated clay particles from the upper horizons.

It is frequently possible to differentiate several horizons of illuviation as it was also possible to distinguish a grading in the horizons of eluviation. Where this occurs it is customary to qualify the symbol B using a suffix which indicates the nature of the enrichment. Thus a horizon enriched with illuviated clay is a Bt horizon, translocated organic matter a Bh horizon and one containing an accumulation of sesquioxides a Bfe horizon.

The soil parent material, unchanged by pedochemical factors, is known as the C horizon and some workers identify a D horizon as being the underlying stratum where it is of different origin from the parent material of the superposed soil.

The foregoing presents a highly generalised version of the processes of soil formation occurring under conditions of intense leaching in temperate climates but individual soils show wide variations of profile

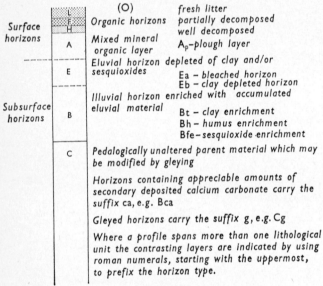

Fig. 8.1   Diagrammatic representation of a soil profile.

which are not always easy to interpret in such general terms. For instance, in the above development of the leaching process it has been tacitly assumed that the parent material remains passive, and that soil formation is essentially a degradation process. This is not necessarily so. A contemporary and continuous rejuvenation of the profile can also be operative in several ways. It is possible for renovation of parent material to occur by the geological erosion of the surface layers which brings the lower horizons within the sphere of biological activity and its ancillary effects. Accumulation of surface material as colluvium or alluvium may introduce fresh material to a site and modify the development processes. By cultivations and tillages man also may check the degradative effects of leaching and exert an influence on the development of the soil profile. Conditions frequently arise to prevent a full and complete leaching of a profile, with the result that the horizon development processes described do not occur or are considerably modified.

The most important of such conditions occurs as a result of water-logging and leads to the production of 'gley' soils, or of 'gley' horizons, within the soil profile. Waterlogging gives an anaerobic environment with a consequent tendency for the formation of minerals in lower oxidative states where possible and blue, grey or green colorations normally associated with ferrous iron compounds are characteristic of gleying. Entire profiles may be so affected or the effects may be limited to zones within the profile where the influences of a high water table permanently operate. In describing soil profiles in which gley features are present, affected horizons have the suffix g attached to the other descriptive symbols.

Other features influencing the formation of horizons may be associated with climatic conditions where surface evaporation of moisture exceeds precipitation. Here soluble products of weathering tend to rise to the surface of the soil, enriching the upper horizons with saline deposits.

Profile Description

The soil profile represents the entire section of the developing soil from the surface downwards and a full description must include a record of all the horizons present including their thickness and the characteristics of the boundaries between them. In Britain certain characteristic horizons are described by symbols, most of which have been given in the preceding text. Below is a summarised list of those commonly found in the soils of that country.

Surface horizons	*Symbol*
Little altered plant debris lying on the soil surface	L
Decomposing organic litter, partially modified	F
Well decomposed organic litter, not mixed with mineral matter	H
Intimately mixed organic matter, predominantly inorganic; the horizon of maximum micro-biological activity, often dark coloured from the humified organic fraction.	A
The ploughed layer of arable soils	Ap
The weakly developed A horizon of young immature soils	(A)

Sub-surface horizons	*Symbol*
The pale coloured sub-surface horizon in certain soils from which clay and sesquioxides have been leached	Ea
A sub-surface horizon in which the clay mineral fraction has been peptised and washed to a greater or lesser degree from the horizon	Eb
The sub-surface horizon of material similar in most respects to the A horizon, the differences being mainly in colour and organic matter content.	(B)
Sub-surface horizons enriched in illuviated clay	Bt
Dark coloured sub-surface horizons enriched in translocated organic matter	Bh
Orange brown sub-surface horizon enriched in translocated sesquioxides of iron and aluminium	Bfe
Sub-stratum of the parent material in which the pedological development of soil horizons has occurred	C
Underlying geological material of different nature to C	D

Horizons that have been modified by the imposition of features associated with anaerobic reduction processes caused by excess moisture (gley characters) have the suffix g tacked on to the primary symbol. Transitional horizons of intermediate character are denoted by the convention A/B or (B)/C and so on.

In American soil classification, the symbols designating horizons differ slightly and are based on those originally suggested by Kubiena i.e. A for surface and eluvial horizons, B for subsurface and illuvial horizons and C for parent material. To these have been added O for organic horizons and R for bedrock.

These master symbols are subdivided by figures and letters so that O1 and O2 are equivalent to the F and H layers of the organic horizons, A1 and A2 are equivalent to the A and E Horizons with A3 representing a transitional stage between A and B horizons. The B symbols are again subdivided according to the degree of illuvial development into B1, B2 and B3. Further subsidiary letter symbols indicate more detailed development features. The important ones are given below.

b—buried soil horizon
ca—accumulation of $CaCO_3$
cs—accumulation of $CaSO_4$
cn—accumulation of concre-
    tionary modules
f—frozen horizon
g—gleyed horizon
h—illuvial humus

ir—illuvial iron
m—indurated horizon
p—ploughed horizon
sa—soluble salt accumulation
si—siliceous cementation
t—illuvial clay
x—fragipan (a brittle sub-surface
    clay accumulation)

All descriptions of soil profiles start from the soil surface, working downwards until the limit of pedological modification of parent material or bedrock is reached.

9 Soils and environment

As has been shown in the previous chapter a natural soil is closely associated with its environment and although the variations that occur between soils in different localities are often progressive and continuous it is possible to recognise dominating features which enable different classes of soils to be defined.

Soil Forming Processes

Soil profiles are described in terms of the horizons that develop within them and which are produced as a result of pedological processes acting on the geologically weathered inorganic parent material. In detail the processes are complicated and have many interactions, but in the first instance they can be separated into three principal pedological developments.

(1) *The organic processes*. The introduction of organic matter into the surface layers initiates the formation of the uppermost soil horizon. The nature of the organic matter, changes that take place within it, and its interactions with the mineral components give rise to different kinds of surface horizons which influence the entire nature of the profiles developing in association with them. In this context the soil microflora and microfauna play a prominent part, as they affect the way in which the organic matter decays and is distributed within the profile.

(2) *The hydrogic processes*. These involve the percolation through the soil profile of water, normally derived from natural rainfall. Not all precipitation necessarily passes through the profile—some may suffer surface run off where the site is on a slope or where the material has low permeability—some evaporates from the surface and some is returned to the atmosphere transpired as water vapour by growing plants.

Any situation has a potential evaporation to set against precipitation and the balance will give what might be termed pedologically active water. If precipitation exceeds evaporation, the excess water, assuming there is no run off, will percolate under the influence of gravity through the soil giving rise to the phenomenon of leaching, and thus will bring lower regions of the profile under the influence of the upper horizons. If evaporation exceeds precipitation, then there is the possibility of an upward movement of water by capillarity and, in consequence, the surface horizons may be modified. These effects may be continuous or intermittent or even alternate.

(3) *Oxidation–reduction processes*. Quite a number of components within the soil profile are susceptible to chemical oxidation or reduction. The majority of soil forming processes are associated with oxidation, atmospheric oxygen acting directly in some few instances to bring about inorganic oxidations and to give an appropriate environment for aerobic organisms to decompose organic matter. If, however, oxygen is excluded either from the whole profile or from particular horizons within the profile, e.g. by waterlogging, then anaerobic organisms develop and introduce reducing conditions which give quite characteristic effects, primarily affecting the iron compounds within the soil. These are reduced to the ferrous state and introduce grey and green colours to the inorganic fabric of the soil; the normal oxidised colours of the inorganic components are shades of brown characteristically associated with ferric oxides. Pedological processes involving reduction of iron compounds in this fashion are referred to as gleying. Where gleying is a standard condition soil colours are often uniformly grey or green—where anaerobic conditions alternate with aerobic conditions ochrous colours may be superimposed on the gley colours and a mottled appearance may develop. Anaerobism also depresses the rate of organic matter decomposition and may lead to organic matter accumulation, even to the formation of peat deposits.

Soil Forming Factors

The soil forming processes alter the soil parent material giving a variety of modifications. These processes, and any soil is a manifestation of their conjoint action, are influenced however, by a number of variables often referred to as soil forming factors. These can be listed:
 (1) The nature of the soil parent material
 (2) Climate
 (3) Relief
 (4) Time
 (5) Man.
In referring to the soil forming processes a measure of reversibility was implied in each.

In the organic processes, organic matter additions and decompositions occur so that the precise OM status of a soil is an equilibrium state between gain and loss. In the hydrologic processes a net situation exists between leaching and evaporation, and in the oxidation–reduction processes also a resultant situation exists. As all these processes are affected by one or other of the soil forming factors and in particular by climate, soil profile development is dynamic, progressing with time, and eventually reaches a stage when it will be in equilibrium with its environment. If the environmental factors change then the equilibrium will also change; a soil will alter and its profile will change to meet the conditions imposed by the new environment. A soil profile represents the condition of pedological development at the time of examination and does not necessarily imply a permanent state.

Soil developments of this kind give rise eventually to a climax soil which has reached an equilibrium with its environment. Perhaps the easiest way to understand the number of development patterns that can occur is to start with a 'no soil' situation and trace the sequence of related soil forms that can develop under different temperature and humidity regimes.

Raw Soils

In the earliest stages soils are characterised by a lack of humus and are essentially accumulations of physically weathered parent material. Chemical breakdown of primary rock minerals is minimal and scarcely affects the nature of the rock residues. The soil profile, such as it is, may be classed as (A)C, indicating the weak development of an organic horizon resting directly on parent material.

Young soils may persist in this state as a consequence of particular climatic conditions or may exist as a temporary phase of development under any climatic condition, particularly where erosion has removed a pre-existing mature soil exposing underlying parent material to weathering influences.

Erosion sites excepted, raw soils occur typically under arctic or desert conditions giving:

(a) Arctic stony soils; the physically weathered stony materials often assuming some kind of geometrical pattern induced by alternating freezing and thawing conditions.

(b) Desert soils; the absence of water precludes the development of plant life and the chemical weathering of primary minerals.

It is usual to divide young soils into *lithosols* and *regosols* according to the nature of the original rock material. The former are derived from hard rocks which give rise to very coarse and stony textured soil while the latter originate on softer deposits giving rise to fine textured profiles.

Young Soils

When arctic or desert conditions relax, the pedological processes are able to proceed a stage further and the more complete development establishes a surface horizon significantly different from the parent material, in which organic matter has been integrated into the mineral fraction and hydrolytic weathering has modified the rock material to a recognisable extent, although physical weathering processes still dominate the profile. This gives an AC type of profile.

Several typical situations can be represented in the establishment of these young soil profiles and it is possible for them to form a stable component in an environment where climatic effects are weak or where the parent material is very resistant to change.

The AC profiles developing from siliceous rocks are called *rankers* and several forms can be recognised.

The *tundra* is a 'young' soil in which climate is the determining factor in development. Characteristic of cold climates with a short summer season a form of vegetative cover develops to give a source of organic matter which, because of low temperatures and water-logging, due either to a permanently frozen subsoil or an impermeable rock sub-stratum, does not decompose or humify very readily. A typical profile might consist of a very thin peaty layer resting on 15 to 20 cm sticky loam predominantly bluish grey in colour merging into 60 cm or 90 cm of a compact stony layer which may be permanently frozen at the lower levels.

Other climatic rankers are the *alpine ranker* and the *xéro-ranker*. The former are associated with high altitudes which limit the development of vegetation and the latter are found in Central European steppe country or in other semi-arid areas where a limited flora can only develop in the wet season.

Ranker profiles are also to be found under climatic conditions normally associated with other soil types. Usually these occupy small areas within bigger regions of contrasting soil. These non-climatic rankers are not in equilibrium with their environment as the alpine or xéro ranker might be considered to be but are maintained in a partially developed state by factors other than climate. These are topographical rankers developing on slopes of siliceous rock, where constant rejuvenation of the soil by erosion of surface material prevents full profile development, or found at the foot of slopes where colluvial movement of eroded material is constantly adding new unweathered products to the soil surface.

SOIL DEVELOPMENT IN SILICEOUS PARENT MATERIAL

Brown Earths

The organic horizons of raw soils developing on non-calcareous parent material tend to be quite acid in character as the decomposition processes give rise to acidic by-products which cannot be neutralised in situ. However, as mineral weathering becomes more significant the secondary minerals become prominent components of the mineral fabric. They include clay minerals and bases which are adsorbed onto the clays and can act as neutralising agents against developing soil acidity. This allows humification to proceed in the direction of mull humus, and the establishment of an A horizon follows as the microflora and microfauna bring about an intimate integration of the organic and inorganic fractions. Underneath the A horizon the mineral fabric shows essentially the same weathered characters as the upper horizon although no significant organic matter penetration is likely to have occurred. This is the (B) horizon and rests directly on the parent mineral matter. It is referred to as an A(B)C profile and is characteristic of *'brown earth'* soils typically to be found on loams or clays under deciduous forest.

The base status of these soils varies but they are generally slightly acid, the pH and exchangeable calcium usually increasing with depth. Most agricultural soils of Great Britain fall into this group which in many areas can be considered a climax soil in equilibrium with its environment.

The equilibrium is maintained by a recirculation of bases by a grass or deciduous vegetation through the decomposition of the alkaline earth rich litter which maintains the base status of the surface horizons. Alternatively the application of lime and fertilisers by man during his farming of these soils also preserves a high level of bases in the surface soil. The mineral skeleton of the soil is also characterised by the deposition of secondary hydrated ion and aluminium oxides as coating films on the surfaces of the sand silt and clay particles, illite and kaolinite commonly occurring in the latter fraction.

A description of this type of profile would follow the pattern typified by that of the East Keswick Series viz.

Horizons
0–25 cm   Greyish brown firm sandy loam with few stones: weak
    A   angular blocky structure; low humus content; earth-
    worms present; sharp even boundary.

Horizons

25–60 cm B	Reddish brown friable sandy loam with frequent angular fragments of sandstone; weak angular blocky structure; few roots; earthworm channels; merging boundary.
60 cm + C	Reddish brown sandy loam, fragments of rotting sandstone, weak blocky structure; few earthworm channels.

Analytical data for this profile is given below:

	Horizon		
	0–25 cm	25–60 cm	60 cm
Sand %	56	57·5	66
Silt %	29	21	17
Clay %	19	18	14
Loss on Ignition	5	3·5	3
C.E.C. (m.e.%)	8·4	6·1	5·3
Exch. Ca „	3·64	3·15	1·88
Exch. Mg „	0·29	0·43	0·58
Exch. H „	4·2	2·4	2·8
pH	5·8	6·0	6·5
% Base Saturation	50	61	47

Leached Brown Earths (Sols Lessivés)

If in situations in which brown earth soils would normally develop soil porosity is excessive or rainfall above average the base status of the soil will be reduced by leaching and the pH of the profile will be lowered. Under conditions of moderate acidity (pH 5–6) it is possible for the colloidal mineral fraction to become peptised and migrate down the profile under the influence of the percolating water. Movement of clay in this way is slow and physical impedence brings about an enrichment of the lower horizons, giving profiles which contain A, E, Bt and C horizons.

Mechanical analysis data for the Rougement Series illustrates this pattern of clay distribution.

Horizon	Sand	Silt	Clay
A 0–10 cm	64	20	16
A/Eb 10–20 cm	69	20	11
Eb 20–46 cm	72	20	8
Bt 46–120 cm	71	5	24
C 120 cm	87	10	3

The clay enriched horizons of leached brown soils often show higher sesquioxide ratio patterns also. This is not necessarily associated with the independent migration of iron and aluminium oxides as most of the clay minerals involved are well coated with films of these compounds which may indeed have an influence on their potential for migration enabling them to become peptised more easily.

Podsolic Soils

Should the base status of a soil become very depleted as a result of much greater rainfall or in cases where the base exchange capacity is naturally very low, as would occur in predominantly sandy situations, soil acidity becomes very high. This has a marked effect on soil development and gives rise to podsolised soils or *podsols*.

The development of soil acidity has two important consequences. The natural vegetative cover changes from mainly deciduous to a heath or coniferous type, and the microbiological population becomes modified, fungal organisms thriving at the expense of bacteria, and less able to bring about effective humification. As heath or coniferous litter is by its nature resistant to oxidative decomposition, this coupled with the changed microflora leads to a change in the nature of the surface humus and an organic horizon of mor humus often differentiated into L.F. and H layers, develops on the soil surface.

This mor humus horizon is usually very acid (pH 4 to 5) and rain water percolating through carries with it the acid products of the decomposing litter which modify the sub-surface horizons. The principle effect is naturally confined to the mineral matter immediately underlying the organic layer and takes the form of a dissolution of the free amorphous sesquioxide coatings of the mineral particles leaving them light-coloured or pale grey, to give a *bleached* horizon. The name podsol is derived from the Russian word for ash or ash-coloured and refers to the presence of this light coloured horizon. The clay content of this horizon is also very depleted.

Beneath the bleached horizon occur one or more illuvial horizons in which the products eluviated through the soil have accumulated. There may be a B horizon notably enriched with an iron rich organic matter, together with one in which migrated iron and aluminium oxides have been redeposited without organic matter, or in some cases the humus rich horizon may not have developed and only a zone of iron enrichment may be found. The B horizons also contain a slight increase in clay content the dominant clay mineral being kaolinite. Occasionally the iron enriched B horizons may be indurated to give a hard pan or may contain hard concretions of ferruginous material; these are referred to as Ortstein formations.

The pattern of profile development under conditions of podsolisation is illustrated by the following description of the Anglezarke Series:

Horizons

0–15 cm 0	⎧9 cm L Litter of bracken remains and pine needles ⎨3 cm F Dark brown partly decomposed litter ⎩3 cm H Black mor humus containing bleached sand grains
15–18 cm A	Black humose sandy loam with abundant bleached quartz grains
18–35 cm Ea	Greyish brown friable sandy loam (pinkish grey when dry); many bleached sand grains; slightly stony.
35–38 cm Bh	Dark reddish brown humose sandy loam; compact.
38–43 cm Bfe	Reddish brown hard sandy loam; few sandstone fragments.
43–66 cm B	Yellow brown friable sandy loam; stony, merging into
66 cm + C	Disintegrating yellowish brown sandstone.

Analytical Data

	Horizon				
	A	Ea	Bh	Bfe	B
Sand %	71	75	67·5	70	60
Silt %	9	10	9	9	19
Clay %	10	11	16	16	17
Loss on Ignition	19	4	7·5	5	4
C.E.C. (m.e.%)	13·4	7·7	21·4	16·1	9·8
Exch. Ca ,,	0·7	0·5	0·5	0·5	0·3
Exch. Mg ,,	0·3	0·1	0·1	0·2	0·1
Exch. H ,,	12·2	7·0	20·5	15·1	9·2
pH	3·6	3·7	3·7	3·6	3·6
% Base Saturation	9	9	4	6	6

Soil Development on Calcareous Parent Material

Rendzina

On chalk or limestone parent material the pattern of soil development in humid temperate climates gives a sequence of profiles somewhat different to that on siliceous parent material.

The initiation of soil development again involves the introduction of a biological component into the physically weathered rock, lichens and mosses commonly being early colonisers of exposed rock sites. The detailed development from this stage is influenced very much by whether the rock is hard limestone or soft chalk, deep profiles being formed much more easily on the latter. The base rich nature of the rock enables a more varied and deciduous flora to develop and also encourages the establishment of an active microbiological population which brings about rapid humification of organic residues. Acid byproducts of humification are rapidly neutralised by the limestone with the consequential solution of rock particles. Under these conditions the humus form is mull and it is usually very dark coloured, often black, as is commonly the case when organic matter decomposes under alkaline conditions.

The young soil profile thus formed is known as a *rendzina*. It is an (A)C profile analogous to the *ranker* as formed on siliceous soils. The name rendzina is of Polish origin and was first used by Sibirtsev in 1898, but the etymological derivation is obscure. These soils develop typically under woodland or old grassland with a dark coloured A horizon often containing fragments of limestone which increase in amounts with depth, merging with brashy rock down to the solid limestone. The whole profile is shallow, the full transition from the well developed granular or crumb structured A horizon to the parent material often taking something like 20 cm.

The following description of a rendzina profile gives the salient features of this soil form.

Horizon

0–20 cm A	Dark grey loam with fragments of limestone, friable; well developed crumb structure; moderate organic matter; highly calcareous; merging into
20 cm + C	Brashy limestone

Analytical Data

	Sand	Silt	Clay	$CaCO_3$	Organic Matter	pH
Horizon A	16	42	14	19	9	8·1
Horizon C	4	37	12	46	1	8·3

Calcareous Brown Earth

Progressive leaching of the rendzina soil would be expected to reduce the amount of free calcium carbonate in the surface horizons and to increase the depth of pedological weathering affecting the material beneath the A horizons to depths at which organic matter infusion does not occur. Under these circumstances there is a tendency to develop *Brown Calcareous Soils* which have a well defined B horizon as part of the solum. They are ABC profiles and have many similarities to the *brown earths* previously described. The base status is, however, much higher due to the presence of free calcium carbonate, and the whole profile is likely to be neutral or slightly alkaline due to the limestone fragments which may not be especially evident in the surface horizon. The colour of the organic matter in the A horizon is often somewhat lighter than that found in the rendzina and the mineral fabric of the profile is reddish brown due to the uniformly distributed secondary ferric oxide gel which forms a film coating the mineral grains.

A profile of this type developing on magnesium limestone is described below:

Horizon	
0–25 cm A	Dark reddish brown friable loam; slightly strong with few small fragments of magnesium limestone; sub-angular blocky structure; sharp even boundary.
25–60 cm B	Reddish brown friable loam; stony with fragments of magnesian limestone; merging boundary.
60 cm + C	Brown friable sandy clay loam; stony merging into massive magnesian limestone.

Analysis (per cent)

	Sand	Silt	Clay	CaCo₃	Organic matter	pH
Horizon A	33	31	24	5	7	7·2
Horizon B	38	31	22	7	2	7·4
Horizon C	44	17	26	12	1	8·2

Under more pronounced leaching the upper horizons become slightly acid as the calcium carbonate is progressively dissolved. This can lead to the mobilisation of the clay fraction and the profile develops similar characteristics to the *leached brown earths* as can be found on siliceous parent material.

It can thus be seen that one of the consequences of pedological weathering is to gradually erase the influence of the parent material on the soil profile pattern, and it is possible to find soils with very similar characteristics on a range of parent materials.

SOIL DEVELOPMENT IN SEMI-ARID CLIMATES

Pedological development under semi-arid climates gives rise to significantly different profiles from those developing under humid regimes. Limited leaching ensures that the products of mineral weathering are retained within the profile to a much greater extent and thus influence soil horizon characteristics and there is frequently a restriction imposed on the vegetation by reason of the climate.

The *chernozem* is an example of the type of soil that can be formed as a result of the influence of climate. The soil parent material normally associated with the chernozem is base rich, but not necessarily limestone, which on weathering gives rise to a high degree of calcium saturation throughout the solum. The secondary clay minerals associated with the weathering taking place in the presence of an alkaline earth rich environment are predominantly montmorillonitic. The climate consists of a cold winter with a quick spring thaw which allows the melted snow and ice to saturate the soil profile. This is followed by a warm dry summer so that only a proportion of the winter moisture drains through the profile. Typical vegetation associated with this type of continental climate is 'steppe' grass.

Conditions in the summer are highly conducive to rapid mineralisation of the grass residues in the base saturated environment, which also attracts a rich mixed fauna, ensuring a thorough mixing of the dark coloured humus throughout the profile. Because of the partial leaching of the soil, the surface layers, although base saturated in so far as the clay and humus colloids are concerned, do not contain free calcium carbonate, whereas the lower horizons usually do contain free carbonates which have been deposited as a result of pedological translocation. It is common to find this secondary calcium carbonate existing as white filaments or as concretions filling former worm casts or rodent burrows which are described by the name 'crotovinas'. Profile development is often quite deep, and because of the dark colour of the humus they are often referred to as 'black earths'. A characteristic profile would contain the following horizons:

Horizon	
0–5 cm O	Surface organic matter derived from 'steppe' grass vegetation.
5–60 cm A	Black humus horizon containing 5 to 12 per cent organic matter, decreasing with depth; well granulated structure with calcium saturated organo-mineral complexes dominating.
60–120 cm Ca	Highly calcareous dark coloured horizon with free $CaCO_3$ and $CaSO_4$ deposits in the form of mycelium threads or crotovinas.
120 cm + C	Base rich parent material.

Under slightly dryer conditions chernozem development is limited by a restriction of humification and the so called *Chestnut Earth* profile develops whereas, under slightly wetter conditions *Regurs*, the *Black Cotton Soils*, are formed where more tropical temperatures also prevail. These have montmorillonitic or expanding lattice clays predominant in the clay complex and often show the characteristics of bad drainage.

ARID SOILS

In regions of limited rainfall where temperatures are such that surface evaporation of water restricts the normal leaching processes, a restriction aided in many instances by impermeable subsoils, the soil form is strongly influenced by the accumulation in the surface layers of water-soluble salts.

Saline Soils

The association of impeded drainage with high temperatures brings about a situation where products of the chemical weathering of rock mineral matter are confined to areas of their production. The soils formed under such conditions are often characterised by the presence of soluble salts within the profile. Sodium chloride and sodium sulphate are the principal components of the soluble salt complex although magnesium compounds may also be present and carbonates and bicarbonates may occur among the anions. In the dry seasons these soils show a white efflorescence of salts on the surface. Due to the high concentration of neutral salts present the clay component of the mineral soil is highly flocculated, giving the soil a good structure. Soils of this kind are known alternatively as *solonchak* soils (the original Russian name), *white alkali soils* or simply as 'saline' soils.

It is not possible to give a general description of the morphological features of solonchak profiles as they show great variation in kind and origin, varying qualitatively and quantitatively as regards the content of salts and parent material.

Alkali Soils

If the conditions under which solonchak soils develop are changed slightly by a moderate relief of the drainage imperfection, an entirely different soil situation becomes apparent. The saline soils have the cation exchange complex dominated by sodium ions and there is in the profile an excess of neutral sodium salts which keeps the clay fraction in a flocculated condition. If this excess of soluble salts is largely removed by virtue of a better drainage system than exists in the saline soil, and providing the precipitation remains small enough only to dissolve and carry away soluble salts, the residual soil remains sodium saturated and the clay fraction becomes peptised. The deflocculation of clay is encouraged by the development of an alkaline reaction following the formation of free alkali in the profile as a result of the hydrolysis of the sodium clay.

$$Na\text{-clay}+H_2O \rightarrow H\text{-clay}+NaOH;$$
$$2NaOH+CO_2 = Na_2CO_3+H_2O.$$

Soils of this kind, with sodium carbonate present in the profile as a secondary product of hydrolysis, are *black alkali soils* or *solonetz*. The most obvious features of solonetz soils are their black colour due to the alkaline solution of humic matter and the deflocculated state of the clay, giving a structureless soil which dries out in the non-rainy seasons and cracks into columnar elements.

Soloti Soils

The saline soils merge into the alkali soils forming a continuous series which can be correlated with precipitation and freedom of drainage. If, in extension of this series, the intensity of leaching imposed upon an alkali soil is increased, the excess of sodium carbonate in the alkali soil will be washed away and the hydrolysis equilibrium of the sodium clay will move towards an increase in concentration of hydrogen clay and the pH of the surface horizons becomes reduced until it drops within the acid range. Under these conditions decomposition of iron compounds introduces a brown coloration to the soil and further leaching of the acid surface layers may cause a translocation of iron with a re-deposition as it reaches less acid horizons. This degradation of the solonetz profile is called solodisation and gives rise to the *soloti* soil.

Lateritic Soils

A feature of the pedalfer developing under temperate climatic conditions is its colour which, making allowance for modifications introduced by organic matter, usually has a marked yellow-brown component. This dominant colour is associated with the weathering products of the iron minerals of the soil parent material which are commonly hydrated forms of iron oxide, including *goethite* ($FeHO_2$) and *limonite* ($Fe_2O_3.n\ H_2O$).

Under tropical conditions, where temperatures rise to a higher level, the soils frequently tend to assume red colours, such as are normally associated with non-hydrated forms of iron oxides, such as *haematite* (Fe_2O_3).

These red soils of the tropics are usually referred to as lateritic soils, a generalised term covering laterites, lateritic loams, and various sub-types and about which there seems to be no general agreement.

The extreme type of laterite has a surface layer notably rich in oxides of iron, aluminium, manganese and titanium, Fe_2O_3 predominating, which collectively may constitute up to over 90 per cent of the mineral matter. The silica content is abnormally low, dropping perhaps to below 2 per cent. Surface horizons of this general composition may originate from igneous parent material having about 7 or 8 per cent Fe_2O_3 and around 40 per cent SiO_2.

The genesis of such soils is not clear although there is a general opinion that the sesquioxide rich surface layers of soil are the result of a desilicification of the parent material followed by a subsequent removal of silica. It is plausible to postulate a rock-weathering environment alkaline in character as a result of mineral hydrolysis, occurring as described in Chapter 2, without the organic matter decomposition associated with more temperate climatic zones to create acidity. Under such alkaline weathering conditions silica could become dissolved to form soluble alkali silicates and be leached from the profile. Organic matter decomposition under conditions of tropical temperatures and suitable moisture is very rapid and complete, breakdown to the ultimate oxidation products, carbon dioxide and water, occurring without the acidic intermediate compounds having anything but a transient existence.

One important purpose of a scientific discipline is to organise available knowledge in a systematic manner. Classifications of any kind are designed to organise the formal knowledge of the objects concerned in such a manner that the properties of individuals and groups can be better understood, and that relationships between groups can be recognised in so far as defined criteria are concerned. Classifications do not add to knowledge and can never be better than the existing state of knowledge, but they can draw attention to gaps in knowledge and can often point out profitable lines of enquiry.

It follows that as knowledge and appreciation of the objects increases classification systems may also need to be changed or reorganised. They can never be absolute, and in practice must be linked to the specific use to be made of them, so in many cases several systems may be in use at the same time for the same objects. Soil as represented by the soil profile is a natural object; and whilst no two individual profiles are precisely alike, groupings of profiles are possible to enable a more facile interpretation of their properties and a better appreciation of causal relationships. Classifications of them may be made to serve many purposes. Their usefulness for agriculture is often the basis of classification, but the criteria to be considered in this context often involves 'non-soil' properties such as climate, topography and economic factors. Non-agricultural classifications of a subjective character are often drawn up for civil engineering purposes involving such things as highway development or suburban expansion schemes but these again often involve a usage component which restricts their general purpose usefulness.

Such classifications may be termed 'technical' as they are organised for some usually narrow purpose and the criteria of grouping are pointed specifically to one target. If, however, the criteria of grouping are based on intrinsic properties not specifically linked to the objective it is hoped to achieve they may be called 'natural'. For the broadest use 'natural' systems of classification are likely to be more generally helpful and therefore should be based on fundamental properties of the objects in a natural state.

Groups within a classification will be based on classes defined by having properties in common. The major classes will have the largest number of similar properties and these will be divided into subgroups as differences between the items increase, or in other words, as the number of shared properties become fewer and fewer. The formal divisions of difference are matters for judgement and may differ between classifications made for different purposes even though they are based on the same properties. In the lowest orders of the system it may be legitimate to include contrived or artefact properties as a basis of differentiation and these may be derived for some technical purpose not easily interpreted from the natural properties.

Objective classification need not in itself give any information about soils, but should enable such knowledge as is available concerning soils to be easily and quickly applied to a particular case if its position in the classification scheme can be ascertained. An effective classification scheme must therefore be based on characteristics of the soil itself and not on environmental features or genetic factors, unless these can be directly and invariably correlated with the soil form.

It must be admitted that a completely satisfactory system covering all kinds of soil has not yet been attained; a good classification can only be built as knowledge of a large number of individual soils is accumulated and, to be effective, it must be able to accommodate all existing individuals. As it is now generally accepted that the soil profile represents the natural form of any soil most modern systems use this as the basis of consideration. The purely utilitarian systems based on agricultural use or engineering properties are scientifically unsatisfactory. Systems based on mechanical or chemical composition are not sufficient to define soils owing to the environmental influences which affect the overall properties, nor is the environment itself sufficient to give a fully effective basis for classification; although if one accepts the argument that a soil is, at any particular instant, the product of its environment impressed upon a variable basic material, then some interpretation of environment must figure in soil definition.

A useful first stage in the general grouping of soils was that suggested by Kubiena and based on the general nature of the horizons present in the profile (see p. 92).

Thus soils can be:

(1) A B C soils. These have an upper leached humus horizon with a definite horizon or horizons of enrichment over the mineral parent material.

(2) A (B) C soils (A B bracket C). Soils with an apparent B horizon below the uppermost A horizon which is not an enriched zone

but merely a region of oxidative weathering above the parent material.

(3) B/A B C soils. A surface enrichment of translocated material as a result of capillary rise characterises this group.

(4) A C soils. The humus-rich surface horizon rests directly on the parent material with no intermediate horizon.

(5) (A) C soils (A bracket C). Skeletal soils of physically weathered rock material with no definite humus layer.

Early attempts to group soils on a petrographical-geological basis were soon found, when extended to cover wider geographical areas,

TABLE 11
SOIL CLASSIFICATION ACCORDING TO CLIMATE
(after Vilensky)

	Polar	Cold	Temperate	Sub-tropical	Tropical
Arid	Tundra	Dry peat soils	Grey earths	—	Red soils
Semi-arid	—	—	Chestnut earths	Yellow soils	Red earths
Moderate	Peat and meadow soils	Black meadow soils	Chernozem	Yellow earths	Laterite
Humid	Podsolised peat soils	Podsolised soils	Podsolised soils	Podsolised yellow earths	Podsolised red earths

to be of little value. Within limited regions of uniform climate and topography it is possible to obtain correlation between residual soils and their source material, providing chemical weathering has not proceeded to too advanced a state. This has been illustrated by Hart in a study of soils of north-eastern Scotland where acid igneous rocks give rise to podsolised soils in sharp contrast to brown earth soils which develop from basic igneous source material. In the case of soils which have reached an advanced stage of maturity, differences in composition of parent material become less evident in the profiles which reflect more and more the influences impressed upon them by pedochemical weathering.

That climate is a factor in the development of distinct profile types was first recognised by the Russian school of pedologists, the founder of which was V. V. Dokuchaiev. He and his successors devised and developed genetic systems of soil classification based largely on climatic belts which could fairly readily be defined and correlated with the profile forms developing in them. The conception of zonal soils was introduced to cover types in which the influence of

TABLE 12

SOIL CLASSIFICATION BASED ON PROFILE LEACHING
(after Robinson)

Degree of leaching			Type of profiles constituting the major soil groups or sub-groups
	Raw humus present		Humus podsols Iron podsols
Fully leached soils (pedalfers)	Raw humus absent	Temperate	Brown earths Degraded chernozem Prairie soils
		Tropical	Yellow podsolic soils Red podsolic soils Tropical red loams Ferrallites
Incompletely leached soils with free drainage (pedocals)	Calcium carbonate present	Temperate	Chernozems Chestnut soils Brown desert soils
		Tropical	Grey desert soils
Soils with impeded drainage (gleys)	Soluble salts absent	Sub-arctic Temperate ,, ,, Sub-tropical and-tropical	Tundra Gley soils Gley podsols Peat podsols Peat soils Vlei soils
	Soluble salts present		Saline soils Alkaline soils Soloti soils

the parent material has been effaced by the development processes, with a secondary division of intrazonal soils, for soils rich in calcium carbonate and those in which lack of free drainage causes modifications of the simple zonal forms. A third group of azonal soils was recognised, in which the composition of the parent material still influenced the nature of the profile. The earlier Russian classification systems are perhaps best summarised by the work of Vilensky who classified the principal soils of the world on a climatic basis, using temperature and rainfall as his principal indices. Table 11 gives a much condensed summary of his classification.

The swing of emphasis from the original geological classifications of soil to climatological ones, whilst drawing attention to the important influence which climate has to play in soil development, does not give a fully satisfactory basis of definition, as climate and soil form are not invariably correlated.

Further approaches to the question are illustrated by the ideas of Marbut in America and Robinson in the British Isles, who endeavoured to combine the chemical and physical nature of soil material with temperature and rainfall influences as modified by drainage conditions. The classification scheme suggested by Robinson, based on type of leaching to which the profile has been subjected, shows this type of relationship (Table 12).

Some schools of thought place an initial emphasis on the overall moisture status of the soil profile, with the form of humus giving a basis for early differentiation. Thus the major soil groups are defined in terms of their drainage status and associated humus form (which is the resultant of moisture, temperature, flora and base status relationships). The major soil groups are broken into sub-groups which are defined in terms of morphological features of the soil profile. The scheme presented below gives a classification of British soils as suggested by B. W. Avery.

A. DIVISION OF AUTOMORPHIC (TERRESTRIAL) SOILS
(Well-drained and moderately well-drained upland soils)

MAJOR SOIL GROUP   SUB-GROUP

I.   *Raw Soils*

Raw, physically weathered soils with feebly developed humus horizons.

(a) On hard rocks or coarse detritus.

(b) On unconsolidated materials mainly blown sand.

1. Montane raw soil.
2. Skeletal Soil.
3. Raw sand.

II.   *Montane Humus Soils*

Physically weathered, mainly grassland, soils of higher mountains with moder-like (non-peaty) humus.

(a) AC or weak A(B) profile on hard silicate rock.

4. Montane humus soil.

MAJOR SOIL GROUP		SUB-GROUP

III. *Calcareous Soils*

Neutral to alkaline forest, grassland and cultivated soils with 'calcmull' or rendzina-moder-like humus, developed on calcareous materials.

(a) AC or weak A(B) profile on highly calcareous rocks.

(b) A(B) profile on limestone or calcareous sediments.

5. Rendzina.
6. Para-rendzina.
7. Red and brown calcareous soils.

IV. *Leached Mull Soils (brown earths and related soils)*

Neutral to moderately acid forest, grassland and cultivated soils with mull or *non-dystrophic moder* humus.

(a) AC or weak A(B) profile.
  (i) On hard rocks.
  (ii) On soft, argillaceous rocks.

8. Skeletal brown earth.
9. Fine-textured regosol.

(b) *Braunerde-like* A(B) profile.
  (i) On hard rocks or non-calcareous siliceous sediments.
  (ii) On quartzose sands and sandstones.
  (iii) On ironstone and related drifts.

10. Normal brown earth.
11. Sandy brown earth.
12. Ferritic brown earth.

(c) With 'textural B' or fossil or relic (B) horizon.
  (i) On chemically weathered residual or transported materials with *braunlehm* formation in subsoil.
  (ii) On argillaceous 'red beds' and related drifts.
  (iii) On sedimentary materials low in iron and rich in montmorillonite.

13. Leached brown soil.
14. Leached red malm.
15. Leached grey malm.

(d) Regraded podsols with mull-like humus.

16. Regraded podsol.

V. *Podsolised (Mor) Soils.*

Strongly acid heath and forest soils with raw or

(a) AC or weak AB profile.

17. Skeletal mor soil.

MAJOR SOIL GROUP	SUB-GROUP

dystrophic moder humus and B horizons enriched in iron, humus or both.

(b) AB profile with A₂ horizon thin and discontinuous or masked by humus.

18. Podsolic brown earth.

(c) With distinct continuous A₂ horizon.

19. Humus-podsol.
20. Iron-podsol.

B. DIVISION OF HYDROMORPHIC (SEMI-TERRESTRIAL) SOIL

VI.   MAJOR SOIL GROUP

SUB-GROUP

Warp Soils (*alluvial soils*).

Natural or man-made soils on recent alluvium, with little or no gleying in upper 40 cm, often subject to floods.

(a) Coarse-textured (sandy, gravelly) soils derived from non-calcareous rocks.

21. Coarse-textured warp soil.

(b) Calcareous soils with AC or weak A(B) profile.

22. Rendziniform warp soil.

(c) Chemically weathered, mainly brown, soils with A(B) profile.

23. Brown warp soil.

VII.   *Grey Hydromorphic*

Periodically waterlogged forest, grassland and cultivated soils with mull or *non-dystrophic* moder humus, and gleyed subsoil horizons.

(a) Medium to heavy-textured upland soils with impeded drainage. (Excess surface water.)

24. Calc-gley.
25. Leached gley.

(b) Sandy soils affected by a high water-table or by lateral seepage.

26. Sandy gley.

(c) Rendzina-like soils affected by a high water-table or by lateral seepage.

27. Ground-water rendzina.

(d) Medium to heavy-textured soils of river flats.

28. Meadow-gley.

(e) Soils of coastal marshes.

29. Marsh-gley.

(f) Regraded gley-podsolic soils with mull-like humus.

30. Regraded gley-podsol.

MAJOR SOIL GROUP		SUB-GROUP
VIII. *Gley Podsolic Soils*		
Periodically waterlogged, strongly acid heath, moorland, and forest soils with raw humus, dystrophic moder, or thin (40 cm) peaty humus formations, and bleached, A_2 horizons, more or less masked by humus.	(a) Podsolised heath and forest soils with gleying or pan formation in B horizon.	31. Gley-podsol.
	(b) Moorland soils with peaty humus, gleyed or moist A_2 horizon, and ungleyed B/C horizons, normally with thin pan.	32. Peaty (gleyed) podsol.
	(c) Moorland soils with peaty humus and gleyed A and B horizons.	33. Peaty podsolic gley.
IX. *Peaty (Anmoorlike) Soils*		
Alkaline to moderately acid dark-coloured soils without (B) horizons, rich in largely decomposed organic matter, formed under waterlogged conditions, with *anmoor,* peatmull, or *peat moder* humus forms.	(a) With strongly gleyed mineral subsoil at ca. 40 cm or less.	34. Fenny gley. 35. Peaty gley.
	(b) With peaty subsoil, mainly derived from basin peat.	36. (Drained) Fen soil.
X. *Peat (Bog) Soils (moss, bog).*		
Acid soils composed largely of partially decomposed plant materials accumulated under waterlogged conditions (including partially decomposed drained forms).	(a) Peat accumulated under the influence of rainwater.	37. Blanket bog. 38. Raised moss.
	(b) Peat accumulated under the influence of seepage from base-deficient rocks or soils.	39. Valley (flush) bog.

One of the biggest difficulties in field soil classification yet to be overcome is in the field of terminology. Several national schemes have been developed through the efforts of research workers in the various countries of the world and although there can be discerned a pattern of similarity it is obscured by the absence of a generally accepted terminology. Kubiena (*Soils of Europe*) has endeavoured to rationalise soil description and nomenclature for European soils, but no general agreement has yet been reached.

The Seventh Approximation

The United States Department of Agriculture Soil Survey has applied itself over many years to construct an internationally valid classification system. Their ideas have developed through many stages culminating in what is now known as the 'seventh approximation'. This is not claimed to be a final version but is felt to provide a reasonable working system. It is a natural system based upon the properties of the soil as studied in the field. It divides soils into Orders, Sub-orders, Great Groups, Sub-groups, Families, Series, and Phases.

The high categories are based upon more generalised similarities and only in the lower categories are differences in detail considered. In the Soil Orders, the differentiae tend to give a broad climatic grouping and these are divided using criteria bringing out the greatest genetic homogeneity in the Sub-orders.

Within each Sub-order, Great Groups are defined using the presence or absence of diagnostic horizons or arrangements of horizons. Sub-groups represent ill defined Great Groups—ill definition covering intergrade soils between different Orders as perhaps reflected by a merging of properties or the presence of a-typical or intermittent horizons.

Families are differentiated within the Groups primarily on the basis of properties of a relatively permanent nature which are important to plant growth such as soil/air, soil/water, plant root relationships or nutrient supplying power. Soil Series represent soil individuals essentially uniform in all profile characteristics, and within a series Phases relate to differences in a single feature such as depth, colour or texture.

So as to increase the international usefulness and to avoid the confusions that might arise if current nomenclatures were used and perhaps yet again redefined, a completely new vocabulary has been constructed. This vocabulary is based mainly on commonly used Greek or Latin roots with syllables linked together in a formalised fashion so that the final names are descriptive in character and indicative of their position in the system.

Soil Orders

There are ten defined soil orders, the names of which all end in -sol connected by either the vowel i or o to a prefix syllable which is descriptive in character.

The Entisols, descriptive mnemonic *recent*, have no developed genetic horizons and include the regosols, lithosols and alluvial soils of other systems. The Vertisols, mnemonic in*vert*, having expanding clays prominently figuring in their mechanical analysis which can cause mass disturbance and distorted movements within the profile. They include, amongst others, the grumusols, regur and black cotton soils.

The Inceptisols, mnemonic *incep*tion, are soils in which horizon development is not well marked, they are young soils and this order would encompass brown forest soils and brown earths.

Aridisols (arid) are soils of dry places and include desert soils, solonchak and some brown earths.

The Mollisols, mnemonic *moll*ify, include chernozems, rendzinas and prairie soils and are characterised by a well developed base saturated mull humus surface horizon with high micro- and macro-biological activity.

Spodosols are allied to the podsols including the iron and iron humus podsols on sandy parent material.

The Alfisols (pedalfer) are of much higher base status although normally found under moist regimes. Grey-brown podsolic soils and planosols exemplify this group.

The Ultisols (mnemonic *ulti*mate) are humid tropical soils and include red-yellow podsolic soils humic gleys and some lateritic soils.

The Oxisols (oxide) are the laterites with free oxides in the surface horizons.

The Histosols (*histo*logy) are the organic soils, the peats and bog soils in which histologically recognisable features of original plant remains are present.

Sub-Orders

The names of the sub-orders are obtained by adding a prefix syllable to a 'key' syllable taken from the Order name. The prefix syllable is mnemonically descriptive of the criteria of division. The key syllables are shown underlined in the order names as follows: *Ent*isol, *Vert*isol, In*cept*isol, Ar*id*isol, M*oll*isol, Sp*od*osol, *Alf*isol *Ult*isol, *Ox*isive, Hi*sto*sol.

The prefix syllables are drawn as appropriate from the following list which also gives the root word, its mnemonic and connotation.

Sub-order element	Derivation	Mnemonic	Connotation
acr	akros (highest)	Acrobat	most strongly weathered
alb	albus (white)	albino	bleached horizon
alt	altus (high)	altitude	cool—high altitude
and			volcanic allophane
aqu	aqua (water)	aqueous	associated with wetness
arg	argilla (white clay)	argillaceous	illuvial clay horizon
ferr	ferrum (iron)	ferruginous	presence of iron
hum	humus (earth)	humus	organic matter
ochr	ochros (pale)	ochre	pale coloured
orth	orthos (true)	orthodox	common type
psamm	psammos (sand)		sandy
rend		rendzina	rendziniform
ud	usus (humid)	udometer	humid climate
umbr	umbra (shade)	umbrella	dark coloured
ust	ustus (burnt)	combustion	dry hot climates

Great Groups

The names are coined by prefixing the sub-order names by an additional descriptive syllable.

Prefix	Derivation	Mnemonic	Connotation
agr	ager (field)	agric	agric
alb	albus (white)	albino	bleached horizon
anthr	anthropos (man)		man made
arg	argilla (white clay)	argillaceous	clay horizon
brun	brunus (brown)		brown
calc	calx (chalk)	calcereous	calcareous
camb	cambiare (to exchange)		alteration products
crust	crusta (crust)		surface crust
cry	kryos (cold)	cryoscope	cold
crypt	kryptos (hidden)		deep horizon
dur	durus (hard)		duripan
dystr	dystrophic (infertile)		low base soln
eutr	eutrophic (fertile)		high base soln
ferr	ferrum (iron)		ferruginous
frag	fragilus (brittle)		fragipan
gloss	glossa (tongued)	glossary	tongued
grum	grumus (crumb)		granular
hal	hals (salt)		salty
halp	haplons (simple)	minimum horizon	
hum	humus (earth)		humic
hydr	hydor (water)		wet
maz	maza (flat cake)		massive
natr	natrium		sodium
ochr	ochros (pale)		pale
orth	orthos (true)		typical
phan		modified from allophane	
plag	plaggen (sod)		man made surface
psamm	psammos (sand)		sandy
quarz	quarz		quartz

Prefix	Derivation	Mnemonic	Connotation
rhod	rhodon (rose)	rhododendron	dark red
sal	sal (salt)	saline	saline
therm	thermos (hot)		hot climate
typ	(typical)		typical
ult	ultimus (last)		strongly weathered
umbr	umbra (shade)		dark coloured
ust	ustrus (burnt)		hot and dry
verm	vermes (worm)	vermiform	wormy-mixed by animals

Family and Series names cannot easily be formalised in any way similar to the higher orders. The number of groups involved is much too large. Soil Series are usually given place names using the locality in which the series was first recognised to identify it. Families are then named after the best known series they contain.

A soil to be classified would first be assigned to an order according to the appropriate general characteristic properties. For example, a soil with diagnostic horizons not representing any form of extreme weathering would fall within the order Inceptisol. Should the profile also be subject to periodic water saturation it would be placed in the sub-order Aquept, and if the mean annual temperature was less than 8·3°C its Great Group would be Cryaquept. Within the Great Groups it would be given a Sub-group class and then a Family name according to the presence or absence of specific profile features as determined by the particular horizons present, all of which in turn determine its Series name as indicated in Chapter 17.

E

In the previous chapters accounts have been given of the soil as a material body, examining its make-up and form without any significant reference to any utilitarian aspect such as may affect its management by man. The properties already described are termed 'pedological' as opposed to the 'edaphological' properties which relate to the soil as a medium in which plants grow. The next chapters will deal with various soil properties which do have a positive bearing on crop growth, but it may be of profit to consider first some general facts concerning plants and get a general picture of the position of soil as it relates to these.

A simple view of the higher plants shows them to possess a root system which is intimately associated with the soil and an aerial vegetative system upon which develop the fruits and seeds of the species. Chemical analysis of plants shows that carbon, hydrogen, oxygen and nitrogen are the major constituents, with potassium, calcium, magnesium, phosphorus and sulphur also present in appreciable quantities. In addition there are traces of many other elements, indispensable for complete and healthy growth, the more obvious being boron, molybdenum, iron, manganese, copper and zinc.

All living plant cells are metabolic systems building up the more complex compounds of their tissues from simpler nutrients. This process demands an energy supply which is met by respiration, a continuous process whereby oxygen is absorbed by the cells and used for oxidative changes with a concomitant release of energy. Carbon dioxide is expelled from the tissue as the end product of the reactions. Certain cells which contain chlorophyll are also capable of absorbing energy from visible light waves of wavelengths between 650 and 700 pm. This energy is used in photosynthetic reactions in which carbon dioxide of the air is absorbed and converted primarily into sugars, eliminating oxygen as a by-product. From the carbohydrates produced by photosynthesis the multifarious compounds comprising the plant make-up are synthesised by complex enzymatic reactions involving the various other essential elements and activated by the energy of respiration.

There are therefore two fundamental types of living cell necessary for a green plant: those in which carbohydrates are produced by photosynthesis to provide the basic organic material of the plant, and those which can absorb water and dissolved mineral salts for the further synthetic reactions which take place in the plant. Many other cells are necessarily present in the highly organised higher plants, to provide conducting tissues, food storing organs, reproductive systems, and so forth.

Of the nutrient elements necessary for crop production, the carbon and part of the oxygen are derived from the atmosphere by the aerial tissues of the plant, while the remainder are absorbed by the piliferous cells of the roots from the growing medium, although in the case of leguminous plants atmospheric nitrogen is indirectly utilised through the organisms living in symbiotic association with the roots.

The soil is involved, as it is the natural growing medium in which the root systems develop, and the fertility of soil is linked with this association. Defining fertility as the capacity to produce healthy plants it is fairly obvious that again we must consider the resultant of many influences, some tending to stimulate and increase growth, others tending to depress development; some of the influences will be inherent in the soil itself, and some will be submerged by overriding climatic effects. Management of soil conditions by man will help to modify some of these fertility factors although many will be outside man's capacity to change them.

For a general survey of fertility factors it is convenient to group them into (a) those inherent in the soil itself and (b) those dependent upon soil condition and thereby subject to a certain measure of control.

Inherent Fertility Factors

The requirements of plant roots for optimum development are that a good and firm anchorage should be provided for the plant, that an adequate and balanced supply of water, air and inorganic nutrients be available, and that there should be no impediment to their growth, either physical or biological.

Soil texture (see Chapter 3) has a bearing on these requirements, influencing all aspects of root development. A single grain sandy texture will not easily provide a firm anchorage and, especially in the seedling stage of the plant, the surface may be moved by winds, causing serious damage. On the other hand, a compact heavy soil may offer physical impediment to root growth and thus reduce growth potential. Texture too influences the moisture status of the soil and the soil air composition and has a bearing on soil temperature.

The depth of the soil profile is another inherent feature which influences fertility, deeper soils enlarging the territory through which the roots can spread and increasing the potential water and plant nutrient supplies. The susceptibility of shallow soils to drought in dry periods may limit growth of many kinds of crop plant.

The presence or absence of a water table and its position give yet another variable factor which affects fertility. Associated with texture and water table are the natural drainage properties of the soil, poor drainage leading to an anaerobic environment in which roots may suffer from high carbon dioxide and low oxygen levels, or they may be injured by microbiologically produced compounds such as ferrous iron compounds or excessive quantities of soluble manganese which are harmful to them.

The composition of the parent inorganic material will influence the natural supplies of inorganic nutrient elements and the ability of the soil to retain them, and the organic matter level will be of great importance on a variety of counts. Humus with its high moisture-absorbing capacity and its base exchange properties helps to maintain adequate growth conditions, its buffering ability to oppose rapid changes in the pH of the growth medium helps to preserve an evenness of soil reaction, and its composition, as a direct source of organic matter, enables the development and maintenance of a large microbiological population which, being responsible for the plant's nutrient nitrogen, is essential to fertility.

Soil Condition and Fertility

In humid temperate climates the one factor responsible more than any other for reducing crop yields is soil acidity. The reaction of the soil is most important and although within the range of crop plants some can be found that will tolerate any particular reaction the majority of economically valuable plants prefer a near to neutral soil. Soil reaction is usually within the control of the farmer and it is a soil condition a more detailed discussion of which is given in Chapter 14.

The structure of the soil is a feature of importance affecting temperature, moisture status and the soil aeration. On soils with textures militating against good fertility it is often possible to improve matters by appropriate measures. The incorporation of organic matter and/or marl with open sand soils will give better structures in that the single grains will better clump and stick together; the incorporation of organic matter with impervious clays will help to open them up and improve their drainage status.

Deficiencies of nutrients can often be made good by the judicious use of fertilisers and manures. The wise management with respect to choice of appropriate crop and rotation, with the use of suitable cultural methods, can frequently minimise inherent features of a soil which would normally be considered disadvantageous. A detailed discussion of soil management is out of place in this volume, but a brief review of some principal operations employed to maintain and enhance the crop-producing power of soils may be useful.

SOIL TREATMENT PROCESSES

Drainage

Excessive water in a soil automatically involves inadequate aeration, lowering of temperature and limited root range. All these conditions are directly inimical to seed germination and crop growth, and the insufficiency of air and the reduced temperature militate against the decomposition of organic matter. The lowering of the water table by the suitable laying of drains tends to remove these undesirable conditions, and although water is removed from the soil the enhanced root range puts the plant in a position to draw on more water than is available in the undrained condition.

In some of King's famous experiments it was found that the difference of surface temperature between drained and undrained land was sometimes as great as 8·5°C.

Autumn Ploughing

Ploughing is the only common cultivation process in which the upper layer of soil is turned over. The effect of this is to bury surface organic matter and incorporate its decomposition products with the soil. The zig-zag surface produced by ploughing facilitates the percolation of winter rain-water and reduces the losses due to run-off and evaporation, which may easily amount to 50 per cent of the rainfall in the climate of Britain. Moreover the exposure of the furrows to alternate wetting and drying out, and to alternate freezing and thawing, increases the formation of soil crumbs and aids the breaking down of the clods of a heavy soil with a consequent improvement of texture and condition.

Spring Ploughing

Soil is sometimes ploughed in the spring to enable the surface layer to dry out slowly and crumble in a condition suitable for making a good seed bed. Also the mulching effect of the furrows is significant at a time when drier weather and greater loss of water by evaporation are likely to set in, and the increased air supply facilitates nitrate production.

Hoeing and Mulching

Whether it be by using an artificial mulch of straw, etc., or by hoeing the surface of the soil, any loose material will reduce the loss of water from the lower layers. Spring ploughing incidentally accomplishes this to some extent, but hoeing is more effective. It has shown that mulching 1 inch deep may reduce the amount of water lost by evaporation from the surface by up to 50 per cent. Deeper mulches are still more effective and a mulch of fine soil is at least as good as or even better than an equivalent thickness of an organic mulch.

Equally important is the aeration of the surface soil and the enhancement of ammonification, and crops growing in spring and summer in properly mulched land are much less dependent on nitrogenous fertilisers than a crop such as winter wheat.

Rolling

When soil is rolled and the particles and their aggregates are pushed together, the conditions become favourable for the retention of water on the surface. In dry weather the water evaporating may involve a considerable loss, but it is frequently necessary to incur the loss for the sake of bringing moisture to germinating seeds and young plants in the surface soil. When the plant roots have penetrated to a greater depth the surface should be hoed and the film broken.

Bare Fallowing

The practice of leaving land uncropped for a season is very long established. To a large extent the facilities for cleaning the land justify the practice, but over and above that there is an evident tendency for a bare fallow to bring about some improvement in the succeeding crop. The reasons for this improvement are not fully understood and may involve water conservation, but it is known that an enhanced production of easily available nitrogen compounds in uncropped land is one important factor.

Addition of Organic Matter

Organic matter is added to the soil in a variety of ways: green manuring (the ploughing in of a crop), application of farmyard manure, and so forth. The consequent addition of humus to the soil has important effects on the soil texture, and fertility is further affected by the stimulus given to the important micro-organic population of the soil. When leguminous crops such as tares and vetches are ploughed in there is the further advantage of enriching the soil in nitrogen. Green manuring tends to reduce the incidence of common scab in potatoes.

Liming

The influence of lime on the soil is very diverse. On heavy soils its action in aggregating the particles and improving the structure is important. Highly unsaturated soils on which sour conditions prevail are brought to a higher degree of saturation, frequently with extraordinary improvements in fertility. The flocculation of clay particles and the increased lime status improve the conditions of micro-biological development and stimulate bacterial activity. This effect is perhaps most apparent on certain grasslands in which bacterial life has been so reduced that a 'mat' of dead grass has accumulated. The proper incorporation of lime in such a surface sometimes brings about the complete destruction of the mat in the course of a very few years. Large applications of quicklime sometimes effect a partial sterilisation of the soil.

Addition of Fertilisers

As has been mentioned, the elements required by growing plants are fairly numerous, but fertilisers containing compounds of nitrogen, phosphorus and potassium are those most commonly used in practice, and those from which increased yields are most likely to arise. The use of these fertilisers is still very largely based on empirical experience and one of the pressing industrial problems which besets soil chemists is the determination of what fertilisers are likely to be most useful, and in what quantities they should be applied to any given soil area.

The following chapters focus attention on those properties of the soil which have a bearing on the capacity of soils to produce crops and may throw some light on the problems involved in bringing about improvements in soil fertility.

Crop growth and development are markedly influenced by temperature, and soil temperature is no less significant than air temperature in this connection. All changes in nature, whether chemical reactions or physical changes of form, and including biological changes which encompass both the chemical and the physical, are controlled by energy considerations. Most changes of these kinds will not proceed at appreciable rates unless the energy levels are adequate, which normally means that the temperature must be sufficiently high, temperature being a measure of energy level. Biological changes are subject also to a control whereby reactions cease at some higher temperature, the reason in this case being associated primarily with an interference with the plant's water metabolism. For any biological process, therefore, there is a maximum and a minimum temperature beyond which development will not proceed, and an optimum temperature which will be within a relatively narrow range at which growth takes place favourably.

Temperature and the Growing Plant

In temperate climates it is usually low temperatures that are responsible for limitations of crop growth and development rather than high temperatures, and there are a number of low-temperature effects collectively responsible for growth restriction.

Within the plant there is a general decrease of metabolic activity as enzymatic reactions are slowed down, associated with a decrease in the permeability of plant cells and an increase in protoplasm viscosity; there is also a retardation of root elongation limiting the ability of the plant to extend its search for water and plant nutrients. The increase in the viscosity of water and decrease in the aqueous vapour pressure of the soil air, linked with a lowering of temperature, make it more difficult for water movement within the soil and there is a decrease in the movement of water from soil to roots.

Temperature and Germination

The ranges of temperature over which seed germination will take place varies somewhat from one plant species to another. The minimum temperature for most common crops is about 5°C and the maximum about 38°C. The optimum ranges for germination of some common British crops are: cereals, 15° to 18°C; brassica, ryegrass, timothy and clovers, 18° to 23°C; mangolds, beets and cocksfoot, around 24°C. Some data published by Haberlandt make clear the important effect of temperature on germination (Table 13).

TABLE 13

NUMBER OF DAYS FOR APPEARANCE OF RADICLE AT DIFFERENT TEMPERATURES

	5°C	10·5°C	15·5°C	18·3°C
Wheat	6	3	2	1·75
Oats	7	3·75	2·75	2
Vetches	6	5	2	2
Sugar Beet	22	9	3·75	3·75

The slower germination at lower temperatures when the seed is in a susceptible stage increases the risk of damage by fungoid and bacterial attack. Rapid germination reduces such risks and a period of good growth in a plant's early stages helps towards an early harvest, which may be a critical factor in those regions where the growing season is of limited duration.

Temperature and Micro-organisms

To a large extent the fertility of a soil is directly related to the development of its micro-organisms. The processes involved in humification (p. 55) are largely the result of bacterial activity, and the decomposition and disappearance or the accumulation of organic matter in the profile is a consequence of the efficiency, high or low, of oxidative organisms. Bacteria are also responsible for the bulk of the nitrogen which becomes available for crop utilisation (p. 137). The development and activity of the microflora react very markedly to temperature variations and the production of nitrate exemplifies this sensitivity. Below 5°C nitrification ceases to all intents and purposes and proceeds at an optimum rate when the temperature is between 25° and 32°C. When the temperature rises to about 95°C complicated effects are introduced as discussed on page 142, which give an insight into the many complexities of dealing with soil problems.

Soil Temperature

The source of heat supply to the soil is primarily the solar radiation, but the responses of soil temperature to the thermal energy received from this source vary widely according to soil conditions, some of which cannot be altered but which, nevertheless, can be taken into account when planning a cropping programme.

Figures obtained at Rothamsted show that the total amount of radiant energy falling on the soil is equivalent to 320 kJ per square centimetre per year, enough to evaporate 127 cm of water, and equal per hectare to the heat derived from burning 1000 tonnes of coal. The amount of heat energy received varies according to the season, ranging from about 168 $J/cm^2/day$ in December to about 1680 in mid-June. Of this total energy received, only about one half of 1 per cent is recovered in the crop.

Variations in Soil Temperature

The amount of radiant heat which falls on a unit area of soil varies according to the time of day, the latitude of the area, the aspect of the site and the atmospheric conditions.

Radiation from the sun comprises rays of all wavelengths from the very short electromagnetic waves through the ultra-violet and visible wavelengths, to the long wavelength rays which constitute the heat rays. The very short wave radiations are absorbed by the outer atmosphere, the longer visible rays and the heat rays easily passing through a clear dry atmosphere to reach the surface of the earth. If the atmosphere contains a high content of water vapour or carries a cloud cover over the earth, much of the thermal energy is absorbed and the amount reaching the ground is greatly reduced. Atmospheric pollution by smoke and industrial fumes has a similar effect. Even with a clear and dry atmosphere there is a certain amount of absorption of the sun's energy and this proportion increases as the passage of the rays through the atmosphere lengthens. This happens when the sun is low on the horizon, when its rays pass through the atmosphere obliquely before reaching a given site. This accounts for the diurnal and seasonal differences and the latitudinal effects on the amounts of heat received by the soil at any one particular spot. Perhaps of a more direct importance to the farmer is the aspect of his land as this gives micro-climatic variations within the regional average and may influence his crop management.

The amount of energy received per unit area decreases as the angle of incidence of the radiant beam increases. Figure 12.1 illustrates similar beams falling on three sites sloping variously towards them. In cases (a) and (c) the same energy is spread over greater areas than in case (b) and its effects per unit area must be correspondingly reduced. Thus, in the northern hemisphere, soils on a south-facing slope actually receive more heat per unit area from the sun than do level areas or soils with a northerly aspect and therefore will warm up more quickly.

Not all the heat falling on the earth's surface is absorbed by the soil

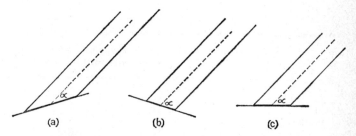

FIG. 12.1 Influence of angle of incidence of radiant beams upon areas affected.

as a certain amount is reflected back into space, depending to an extent upon the colour of the absorbing surface.

It is a well known physical property that dark-coloured bodies absorb heat more efficiently than light-coloured bodies, other things being equal. This effect was investigated as long ago as 1830 when Schübler examined the temperatures of comparable blocks of soil artificially coloured by dusting the surfaces with magnesium carbonate and soot to give white and black surfaces, the temperatures of which he compared with untreated soils. Table 14 gives some of his results, the temperatures being measured in degrees Centigrade.

TABLE 14

Soil	Natural	White	Black
Dark Grey Garden Soil	7·33°C	5·74°C	10·52°C
Loam	6·94	5·65	9·72
Yellow Clay	6·72	5·74	9·84
Light Grey Quartz Sand	6·94	6·20	10·61

In the above table the temperature differences for each soil are due to differences in efficiency of heat absorption. The increase in temperature following the full absorption of a given amount of heat also varies from soil to soil, depending upon the specific heat, which can be defined as the relative amount of heat required to raise the temperature of 1 gramme of the material under consideration through 1°C. The importance of this can be illustrated by comparing the effects of a given quantity of heat on the temperatures of water, quartz sand, kaolinite clay and humus, as shown in Table 15.

TABLE 15

	Specific heat	Increase in temperature on absorption of 4·2 joule
Water	1·0	1°C
Quartz Sand	0·2	5
Kaolinite Clay	0·23	4·3
Humus	0·45	2·2

From the table it can be seen that for any given increment of heat, a sandy soil will react by increasing its temperature to a higher level than clay or humus and the increase will be five times that given by water. It follows therefore that soils with a high water content will respond much less readily to changes in temperature and explains why wet clay soils with their high moisture content are considered 'cold' soils as opposed to 'warm' light soils which do not retain any large proportion of water.

A further difference arises between a wet and a dry soil. The evaporation of water requires the provision of heat, which is taken from the immediate surroundings (the soil) whose temperature is thereby reduced.

Profile Temperatures

Under the influence of the sun it is the surface of the soil that is warmed, and in this respect the presence or absence of vegetative cover is of some significance. Bare soils heat up much more quickly under the influence of solar radiation than do those carrying vegetation by reason of the effect as a blanket to heat transfer. It is, however, also true, when soils are losing heat, that bare soils cool more quickly, and as a result soils under the cover of a crop are cooler in summer and warmer in winter than bare soils.

The heat received at the surface is conducted downwards but not to a very great depth. At depths of about 1 metre the heating effect of the sun is not appreciable. The surface reaches a higher maximum temperature than air temperature, and it reaches its maximum soon after the air maximum temperature has been reached—that is, shortly after midday. The maximum temperatures attained below the surface are lower the greater the depth. Not only are they lower than the maximum at the surface but they are attained later in the day. This vertical variation of temperature is dependent upon the thermal conductivity of the soil which is a product of various factors. The thermal conductivity of soil solids is greater than that of soil water which is in turn greater than that of soil air. The lower specific heats of soil solids and soil air mean that their temperatures will be raised more quickly than soil water, and the facility with which heat passes through the solid particles of the soil is restricted by the small areas of contact between them.

At Rothamsted Experimental Station continuous records of soil temperatures have been published and it would seen that there is a pattern of variation which varies from summer to winter.

Summer Soil Temperatures at 15 cm Depth

There is a marked daily variation in soil temperature during the summer. At a depth of 15 cm the daily rise begins about 9.30 a.m. and continues until about 4.30 p.m. The maximum then reached is maintained for a short time, after which the temperature falls, more slowly than it rose, until about 7.30 a.m.

The mean temperature is passed about midday and midnight. From midday to midnight therefore is the warmest period in the soil.

The maximum temperature at 15 cm is attained about three hours after the maximum temperature in the air. It is usually about the same as the air temperature, and at Rothamsted was about 22°C. (At the surface of the soil the maximum temperature is, of course, higher than the air temperature.)

The minimum temperature at 15 cm is attained about 7.30 a.m. and is 6°–8°C higher than the air minimum, and at Rothamsted was found to be about 18°C.

Winter Soil Temperature at 15 cm Depth

In winter there is no daily variation of soil temperature as there is in summer. A variation may be continuous over a period of several days.

The temperature at 15 cm is never quite so high as the air temperature.

The minimum temperature is generally about 3°C higher than the air minimum, and is reached later in the day in winter than in summer, usually about 10.30 am. The air minimum is attained about the same time all the year round.

The change from winter to summer variations is fairly sharp.

13   Mineral nutrients in soils

(i) SOIL NITROGEN

Nitrogen appears to be normally assimilated by plants in the form of nitrate. There is, however, considerable evidence to show that ammonium salts can be utilised directly as nitrogen sources. There would appear to be no general uptake of amino acids or lower aliphatic amines, but compounds of these types may be variously useful according to plant species, e.g. aspartic acid has been shown to be capable of absorption by pea and clover crops, and glutamic acid by wheat and barley. A wide range of amino acids can be useful for tomatoes, although plant growth in general appears to be depressed in culture media containing too high a concentration of amino acid, viz. of the order of 1,000 parts per million. In soils the free amino acid content is very low, only trace quantities being present.

Within the plant nitrogen forms a characteristic constituent element of proteins, which may constitute up to 20 per cent or more of the dry weight of a plant, as well as being found in many other organic compounds associated with plant life. All the nitrogen naturally absorbed is taken up via the root system from the soil.

The effects of nitrogen on plant growth, as they are empirically observed, are fairly characteristic. Suitable nitrogen compounds tend to increase the bulk of plants, and frequently to increase the bulk of the less valuable part of the plant. Increasing quantities of nitrogen will continue to increase the yield of straw after they have ceased to increase and even after they have depressed both the yield and the quality of the grain. With root crops the tendency is similarly towards a high ratio of leaf to root. The ripening of grain is retarded by too great a proportion of nitrogen in the medium of growth, and the liability to disease is considerably enhanced.

Nitrogen starvation is characterised by a general yellowing of the leaf and stunted growth.

The amounts of nitrogen found in soils vary enormously. Soils rich in organic matter contain up to 1 per cent or more; at the other extreme, infertile sands may contain less than 0·05 per cent. Normal figures for average and moderately fertile loams are from 0·10 per cent to 0·30 per cent. The total amount present, however, is no guide to the amount available for the plant, as the facility with which the nitrogen in its various compounds can be converted into nitrate, or some other easily assimilated combination, is the chief factor which

135

determines the usefulness of the nitrogen. There are many substances of which leather waste is a notable example, which are sometimes suggested for application to the soil on account of their high nitrogen content, but which, although they literally enrich the soil in nitrogen, are almost without value to the plant since the nitrogen compounds are extremely stable and very resistant to the changes shortly to be considered.

The Natural Sources of Soil Nitrogen

Organic Nitrogen

The chief natural source of soil nitrogen is the protein and other nitrogenous bodies contained in plant and animal remains which become incorporated in the soil, and it is chemically mainly α amino nitrogen. The oxidation of organic matter in the soil has been referred to in Chapter 5; the fate of the nitrogen during that oxidation is the special consideration which arises here.

For plant residues the ratio of nitrogen to carbon varies from about 1:40 to 1:25 (the narrower ratio is generally found in legumes). In soil organic matter (humus) the ratio of nitrogen to carbon is usually about 1:10 although variations between one soil and another are a little greater than were once thought.

The decomposition of the organic complex in the soil is brought about by a variety of organisms, chiefly fungi and bacteria, and during the course of the decomposition processes ammonium nitrogen appears to be formed as a first principal intermediate. This stage in the decomposition of the more complex nitrogenous bodies is referred to as ammonification and the ammonia formed provides the major source of available nitrogen for plant utilisation. Of the ammonia nitrogen thus formed, part may be incorporated in humus synthesis, part may combine with the colloidal complex of the soil, part may be utilised directly by plants or incorporated in growing micro-organisms, and part may be oxidised by bacteria to nitrite and nitrate. Of the total ammonia nitrogen produced by the ammonification process the amount available for nitrification, as the further oxidative reactions are called, is low, varying according to conditions from nil to about 30 per cent, and nitrification would appear preferentially to utilise ammonium ions which are associated with the clay complex—in other words, the exchangeable ammonium ions.

Under normal conditions the oxidation of ammonium compounds appears to follow a fairly specific route: *Nitrosomonas* and *Nitro-*

coccus bacteria convert ammonium to nitrite, and the nitrites are further oxidised to nitrate by *Nitrobacter* bacteria. These organisms appear to be specific to their particular stage of the change: only such as *Nitrosomonas* and *Nitrococcus* can effect oxidation of ammonium compounds; and only *Nitrobacter* can effect the oxidation of nitrites.

The general nitrogen requirement of the soil microflora is more readily met by ammonium nitrogen than by nitrate nitrogen (in contrast with the higher plants which prefer a nitrate source); this means that under normal conditions the level of ammoniacal nitrogen is kept

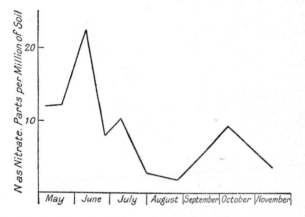

FIG. 13.1 Fluctuation of nitrate content of a Yorkshire (Garforth) soil.

low, being in the order of 2 or 3 parts per million, although the values fluctuate a good deal.

The level of nitrite in soils is usually much lower than that of ammonia, oxidation by *Nitrobacter* being a much quicker process than the oxidation by the first stage nitrifying organisms, the nitrate end product dominating the process.

The amount of nitrate present in the soil at any time is the resultant of the rate of formation and the rate of dissipation which will be discussed below, but measurements of soil nitrate indicate seasonal fluctuations with peak productions occurring in spring and autumn. Figure 13.1 shows this seasonal variation in nitrate content on a Yorkshire soil, the spring maximum giving a nitrate level of around 20 parts per million, dropping to about 2 parts per million in summer, with a second peak figure in the autumn of 10 parts per million. This curve is very similar in trend to curves obtained when micro-biological activity is investigated, the similarity helping to justify the conclusion that microbial activity and nitrate production are related.

Atmospheric Nitrogen

There exist in soils certain bacteria which are capable of taking up nitrogen from the atmosphere. This nitrogen becomes incorporated in the protein of the bodies of the organisms and so enriches the soil in nitrogen. Some of these organisms exist in the soil quite independently of growing plants, but there is one group which, although it exists in the soil, only functions as a nitrogen-fixing organism when living symbiotically with legumes.

Of the nitrogen-fixing bacteria which are independent of plants, the chief are the butyric acid bacillus (*B. amylobacter*) which is commonly called *Clostridium pasteurianum* and *Azotobacter chroococcum*.

The *Clostridium* is a spore-forming organism which fixes nitrogen only in the absence of oxygen; in the soil there are certain bacteria associated with it whose function appears to be to remove oxygen from the neighbourhood. *Clostridium* operates in soils over a wide range of soil reaction, and appears to develop in quite acid soils. Grown in culture media containing sugars, it produces butyric and acetic acids, among other things, during the decomposition of the sugars.

Azotobacter is a non-spore-forming organism and functions aerobically. It is very susceptible to soil acidity and is rarely present in the absence of calcium carbonate. The failure of *Azotobacter* to develop in soils lacking in lime has been used as a test for lime deficiency. The course of decomposition of sugars by *Azotobacter* is apparently different from that by *Clostridium*, as no butyric acid is produced. In culture media, where sugars are used as a source of energy for the organisms, *Azotobacter* fixes four or five times as much nitrogen as *Clostridium* for the same weight of sugar.

While it is known that these two bacteria exist in soils, and that in artificial media they fix nitrogen from the atmosphere, there are yet no reliable data from which any deductions can be made as to the amount of nitrogen fixed by them in the field, and to what extent the farmer is indebted to them.

Another organism, *B. radicicola*, enters leguminous plants from the soil and is responsible for the familiar nodules on the roots of these plants. Living thus in symbiosis with its host plant, this organism fixes considerable quantities of nitrogen. Long before this organism was isolated or its function understood, it had been established by experience that legumes, notably clover, were in some way responsible for a marked improvement in succeeding corn crops. In 1874 Lawes at Rothamsted had actual analytical data to show that in one of his experiments the total barley crop following clover contained over 70 per cent more nitrogen than the total (and smaller) barley crop which was preceded by barley. It was not until 1888 that Hellriegel and Wilfarth demonstrated that legumes, as distinct from other natural orders, took nitrogen from the atmosphere as well as the soil, a discovery which was shortly followed by the isolation of the organism. This discovery was of outstanding importance since it satisfactorily explained the improving effect of a clover ley upon a succeeding corn crop, and it ended a great controversy as to whether plants did, or did not, take nitrogen from the atmosphere.

The atmosphere does make a further small contribution to the soil's nitrogen content in that falling rain and snow carry down small quantities of ammonia and nitrate nitrogen. The amounts are variable with season and locality, ranging from 1 kg of nitrogen per hectare per annum to about 70 kg. The ammoniacal compounds are derived from domestic and industrial fumes, whilst the nitrate is mainly the result of electrical discharge in the atmosphere bringing about the combination of oxygen and nitrogen to form oxides of nitrogen.

THE NATURAL LOSSES OF SOIL NITROGEN

Apart from the removal by crops, there is a variety of ways in which nitrogen may be lost from the soil.

Drainage

Nitrate ions which are formed in the soil as a by-product of micro-biological metabolic processes will be excreted into the soil solution. As there is no other constituent in the soil which has any great affinity for nitrate ions these will be removed as rapidly as the drainage conditions permit. This will apply also to any nitrate ions which may be added to a soil in the form of a fertiliser dressing. On a small plot at Rothamsted which has been kept free from all vegetation, the loss of nitrogen (which in 35 years has exceeded 1,125 kg per hectare) from the top 25 cm is practically the same as the nitrogen which has appeared in the drainage water as nitrate. Rothamsted also provides another striking illustration of the removal of nitrate in drainage water. Among the continuous wheat plots in Broadbalk field there is one which has received no manure since the inception of the experiment in 1843 and which for many years has given an annual average yield of about 450 litres of grain. In Agdell field there is an un-manured plot which grows wheat in alternate years and is bare fallowed in the intervening years. The average yield in those alternate years is approximately 600 litres. The higher yield of wheat following fallow, when compared with wheat following wheat, is attributed to the nitrate produced during the period of fallow, and it is interesting and important to note that in wet seasons, when nitrate is leached, the yield of wheat following fallow is much below the average, and only slightly greater than wheat following wheat. In abnormally dry seasons the difference between the yield of wheat following fallow and of wheat following wheat is most marked.

Loss as Free Nitrogen

When nitrogenous organic matter decomposes under natural conditions there is an evolution of gaseous nitrogen. This has been known for some time to take place in sewage beds, and was demonstrated by Russell and Richards to be one of the chief sources of loss of nitrogen from manure heaps. The chemical changes involved are somewhat obscure, but appear to be associated with a co-existence of oxidation and reduction environments. Such conditions do occur in manure heaps where there is an exclusion of air from certain parts and a circulation of air between those parts.

A possible mode of reaction has been postulated which may operate to some extent and involves nitrite ions which under certain conditions react with amino groups or ammonium salts. Under slightly acid conditions nitrite formed during nitrification could form nitrous acid which might react with these compounds with liberation of gaseous nitrogen.

$$R\text{-}NH_2 + HNO_2 \rightarrow R\text{-}OH + H_2O + N_2$$
amine
$$R\text{-}NH_4 + HNO_2 \rightarrow R\text{-}H + 2H_2O + N_2$$
ammonium salts

Little is known of the extent and importance of this loss of nitrogen from soils but evidence of its existence can be found. Losses of nitrogen have been described from aerated Canadian soils which the meagre drainage is inadequate to explain. There has been an average annual loss of something like 140 kg of nitrogen per hectare from the farmyard manure plot which received 14 tonnes of manure annually on the continuous wheatfield at Rothamsted. This high figure, and the fact that the drain from this field scarcely ever runs, indicates the possibility of loss of gaseous nitrogen in some such way as takes place in sewage beds and manure heaps.

Loss as Volatile Ammonia

In alkaline soils the ammonia produced by degradation of nitrogenous organic matter must, in part at least, form ammonium carbonate, bicarbonate or even hydroxide, and volatilisation of ammonia on the dissociation of these compounds may occur. The potential for such a volatilisation will vary according to a number of factors, which include:

(a) area of soil-air interface, i.e. soil structure;
(b) saturation deficit of air with respect to NH_3 gas;
(c) ammonia vapour tension in soil, which will be proportional to the concentration of ammonium salt present and the pH;
(d) temperature.

Experimental figures for losses by the volatilisation of ammonium salts from soil surfaces vary from less than 5 per cent volatilisation at pH 5 to over 50 per cent at pH 8, although there is no quantitative information as to the importance of this mechanism for nitrogen loss for soils in the field.

Denitrification

There is another possibility of loss of gaseous nitrogen from soils which is not related to the chemical reactions described above. Under anaerobic conditions a large number of soil organisms are capable of decomposing soil nitrate to obtain the oxygen requirements of their metabolic processes. This biological reduction reverses the nitrification process, reducing nitrate to nitrite and nitrite to ammonium or to gaseous nitrogen with a consequent loss of the free nitrogen formed.

Biological Fixation of Nitrogen

When there is in the soil a very high proportion of non-nitrogenous organic matter and the nitrogenous matter falls below a certain percentage, the soil organisms attacking the non-nitrogenous matter, and for which organic nitrogen is now limited, tend to take nitrogen from soil nitrate, thereby bringing that nitrogen into organic combination. For this reason the application of straw, sawdust, etc., to soils tends to reduce the amount of nitrate present. This is of course not strictly a loss of nitrogen, but involves a temporary reduction in the amount which is available to the plant.

The principal nitrogen changes that have been described above may be summarised as shown in Figure 13.2.

Partial Sterilisation

Farming experience long ago discovered that crop growth is often considerably enhanced in places where hedge trimmings, etc., have been burned. Attempts were made to explain this as the effect of the potash which remained after burning such things, but the frequently remarkable increase in the bulk of the crop was rather at variance with the usual effect of potash. Moreover, similar results were obtained in

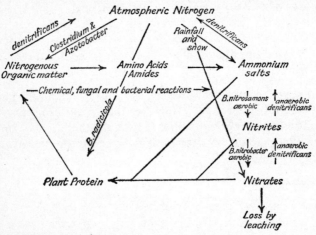

FIG. 13.2 Nitrogen cycle.

many experiments which involved heating the soil or the use of anti-septics, and in which additions of potash did not arise. For example Seton and Stewart, in experiments designed to ascertain whether common scab in potatoes was a mechanical or biological effect, incidentally noticed, but were unable to explain, a much bigger yield of potatoes in soils which had been sterilised with formalin than in unsterilised soil. In one experiment they obtained nearly three times the yield of potatoes as a consequence of treatment of the soil with formalin.

Systematic investigation of this matter was initiated by Russell. Russell and Darbishire investigated in a long series of experiments the effect of heat and of antiseptics on crop yield. With non-leguminous crops they obtained increases varying from 50 per cent to 350 per cent as a consequence of heating the soil to a temperature of 90°–

95°C. By using various volatile antiseptics they also obtained considerable increases in yield. With legumes, however, there was generally no increase consequent upon the treatment. The fact that antiseptics as well as heat have this effect, and the fact that it is, generally speaking, not apparent in legumes (which are able to obtain their nitrogen from the atmosphere), clearly indicates a biological effect connected with the nitrogen nutrition of the plant. In a series of classical experiments, Russell and Hutchinson came to the conclusion that in the soil microflora there exist protozoa which feed upon the bacteria and set a limit to their development, and that the effect of heat and antiseptics is to destroy protozoa but only to suppress, without completely destroying, all the bacteria. After this initial suppression, the bacteria are able to develop unhampered by their former enemy, with a consequent increase in the production of soluble nitrogen. The Rothamsted experiments have shown that after the partial sterilisation treatment there is a reduction in the number of bacteria, followed in a few days by an increase to a level which far exceeds the original number, and that this increase in bacterial population is accompanied by a larger production of ammonia. Moreover, a long series of daily counts showed that whenever the number of protozoa in the soil increases, the number of bacteria is reduced; and that as the protozoa diminish, so the bacterial numbers increase.

It is possible that other effects in addition to the extinction of protozoa play a part in the results which follow partial sterilisation, at any rate by heat treatment. The effect of heating soils involves a number of physical changes which influence the structure of the soil and of chemical changes which render a number of soil constituents, e.g. potassium, more available to the plant, and these effects may also be of some significance.

NITRIFICATION AND FARM ECONOMY

It will be clear from what has been said, that the formation of nitrate in the soil is apt to give rise to considerable loss by drainage. Unless the root hairs of a plant are ready to absorb the nitrate immediately it is formed it will descend with the drainage water. Moreover, the formation of nitrate takes place to a greater extent in uncropped land than in cropped land, and in an average season there is probably more nitrate formation immediately after harvest than immediately before. So far as can be judged from available data something like half the nitrate produced in soils is, on the average, lost by drainage.

In spite of this it has been the invariable orthodox teaching of the agricultural scientist and the scientific agriculturist that nitrification is a wholly desirable process. There are two facts which have probably given rise to this belief. One is that those soil conditions such as adequate aeration and water supply, suitable temperature and reaction, which are conducive to soil fertility are also conducive to nitrite formation. The second is that in experimental work with fertilisers nitrogen in the form of nitrate has generally given higher returns than nitrogen in the form of an ammonium compound. The first of these is obviously no evidence whatsoever that nitrate formation is desirable, and the second fact is irrelevant. If nitrification were stopped and the nitrogen accumulated in ammoniacal combination, the amount which would accumulate as ammonia which is not lost from the soil would be very much greater than the amount present at any one time as nitrate, and while, for equal weights of nitrogen, nitrate may be more efficient than ammonia it is indisputable that several times the weight of nitrogen as ammonia is more effective than a given weight of nitrogen as nitrate.

The adherence to the belief that nitrification is a desirable process is all the more remarkable in view of the experimental results of partial sterilisation of soil, a process which enhances fertility but in which the nitrifying organisms are killed off, with the result that the nitrogen accumulates as ammonia.

(ii) PHOSPHORUS

Farmers have recognised for well over a century that many crop yields can be increased by applying to the soil materials some of which have been shown to owe their beneficial effects to a content of phosphorus. The bone products used originally have been largely supplanted in more modern times by rock phosphates, treated in various ways to increase the solubility of the phosphorus in them and thus increase the efficiency of its utilisation.

All plants require phosphorus for their growth and development in significantly large quantities, it being involved as a constituent element in many specific compounds making up the plant's structure as well as playing an important part in the metabolic processes which enable the plant to develop and complete its natural life cycle.

During growth, phosphorus is continuously being translocated throughout the whole plant and it is found distributed in every part of the plant tissue. It is a constituent element of nucleoproteins which are involved in the cell reproduction process, and it can be stored in the plant tissues as a phospholipoid or as phytic acid or phytic acid salts.

Metabolically, phosphorus is an essential component in the reactions of carbohydrate synthesis and also of carbohydrate degradation, enabling energy liberated during the breakdown of hexose sugars to be preserved in carboxyl phosphate or amino phosphate linkages which, on subsequent hydrolytic breakdown, yield an abnormally high amount of energy for use in general biosynthesis within the plant.

It can easily be shown that the total content of phosphorus in a soil is not necessarily related to its ability to supply a growing crop with the phosphorus it needs, and it can also be demonstrated that the addition to a soil of phosphorus in a form known to be readily available to plants is a wasteful process, the plant only being able to recover a very small fraction of the added nutrient.

Most of the phosphorus absorbed by plants is in the form of an orthophosphate, the dihydrogen orthophosphate ion H_2PO_4' being the most effective carrier. It can, however, be absorbed in other forms, particularly in some organic combinations; it has been demonstrated, for example, that phytic acid and lecithin can act as direct phosphorus suppliers for certain plants and that nucleic acids, which contain phosphorus, can be enzymatically decomposed at root hair surfaces with the formation of inorganic phosphates.

Inorganic Soil Phosphorus

The primary source of soil phosphorus is the apatite group of minerals $3Ca_3(PO_4)_2.CaX_2$, where X_2 may be $(OH)_2$, Cl_2, F_2, or CO_3. Apatite minerals occur as such in soils, being derived directly from the original rock formations. They are all of low solubility and as plant nutrient elements appear to need aqueous solution to facilitate their uptake apatite minerals are considered of no value as direct nutrient compounds. Secondary phosphorus minerals are also to be found in the soil. These are normally orthophosphates of iron, aluminium and calcium. *Vivianite* ($Fe_3(PO_4)_2.8H_2O$) and *dufrenite* ($FePO_4.Fe(OH)_3$) are perhaps the most common iron phosphates; *variscite* ($AlPO_4.2H_2O$) and *wavellite* ($Al_3(OH)_3.(PO_4)_2.5H_2O$) being more common aluminium phosphates. Magnesium phosphates have not been found in soils, although a mixed phosphate, *lazulite* ($Mg.Fe''$)$Al_2(OH)_2(PO_4)_2$, has been shown to occur.

As sources of phosphorus for plants these minerals are not considered to be very useful, although opinions have been offered that vivianite and variscite may be of some value as direct sources of phosphorus.

The calcium phosphates are not so easily defined. Being tribasic, orthophosphoric acid forms three series of salts according to the number of hydrogen atoms replaced in the molecule. The primary, secondary and tertiary calcium salts are monocalcium phosphate $Ca(H_2PO_4)_2$, dicalcium phosphate $CaHPO_4$ and tricalcium phosphate $Ca_3(PO_4)_2$. Of these salts monocalcium phosphate has the highest water solubility, while tricalcium phosphate is practically insoluble, and in an aqueous medium in vitro it is possible for the three salts to exist in equilibrium.

$$Ca(H_2PO_4)_2 \rightleftarrows CaHPO_4 \rightleftarrows Ca_3(PO_4)_2$$

Under acid conditions the equilibrium moves to the left and as the pH increases the more insoluble compounds tend to become dominant. In the environment normally existing in the soil where the pH range is restricted to fairly narrow limits around neutrality, monocalcium phosphate would probably be the principal stable compound. Under more alkaline soil conditions which involve the presence of higher calcium ion concentrations in vitro, phase studies indicate that tricalcium phosphate itself is not likely to exist but that more basic and insoluble phosphates of calcium, such as calcium octaphosphate $Ca_4H(PO_4)_3$ and hydroxyapatite $3Ca_3(PO_4)_2.Ca(OH)_2$, are likely to be present.

Because of the fact that in acid soils soluble phosphate will combine with iron and aluminium to form insoluble compounds, theoretical considerations would suggest that a soil reaction within the approximate limits pH 6 to pH 7 would give a maximum level of inorganic phosphorus compounds having a significant availability.

Organic Phosphorus Compounds

Soil organic matter contains phosphorus, empirical analysis indicating that in a normal brown-earth type of soil about 20 per cent of the total soil phosphorus may be organically combined. Three classes of compound appear to be principally concerned: nucleoproteins, phospholipoids, and phytin derivatives. The nucleoproteins are constituents of the nuclei of biological cells and the bulk of such compounds will be associated with the living microbiological flora. The phosphorus contained in nucleoproteins is not liberated as phosphate by the addition of alkali but acid hydrolysis does liberate phosphorus from the sugar phosphate moiety of the molecule.

The phospholipoids are widely distributed in plant and animal tissues. They are triglycerides analogous to fats in which one of the glycerol hydroxyl groups is esterified to a choline ester of phosphoric acid. On hydrolysis phosphoric acid may be liberated.

Most significant perhaps of soil organic phosphorus compounds are the phytic acid derivatives. Phytic acid is inositol hexaphosphate and has twelve replaceable hydrogen atoms capable of forming salts. The calcium and magnesium salts are very insoluble and resistant to breakdown but may be eventually hydrolysed microbiologically with the liberation of the otherwise very stable phosphate.

Phosphate Fixation

It has been stated that plant recovery of phosphorus supplied in fertiliser form to the soil is rather low, and more than 85 per cent of added phosphorus appears to be irrecoverable. Examination of drainage effluents shows that the loss of fertiliser phosphorus from the soil is not an absolute one as the concentration of phosphorus in drainage water is so small as to be negligible. This effect was noticed as long ago as 1850 by Thomas Way who found that when he poured a solution of a soluble phosphate through a column of soil no phosphate appeared in the filtrate. This ability of soil to retain phosphate ions (commonly referred to as 'phosphate fixation') has received much attention and it is indeed a problem of great economic importance.

Phosphate fixation is not restricted to any one particular soil kind, occurring in all types of soil at all levels of pH. It is a composite reaction and in any particular case one or more of the component reactions may dominate the overall phenomenon. It will be convenient to describe the processes contributing to the whole separately but their relationship must not be overlooked.

In the majority of soils a considerable portion of the fixed phosphorus is held in combination with inorganic constituents. In neutral or alkaline soils calcium and perhaps magnesium are perhaps the principal fixing agents, although iron, aluminium and clay minerals may play some part in the process. In acid soils iron, aluminium and possibly the clay minerals are the more important agents involved. This inorganic fixation of phosphorus takes place via mechanisms involving chemical precipitations, anion exchange or surface adsorptions.

Additional to inorganic fixation processes, organic reactions may be responsible for the locking up of some of the available nutrient and may be classed together as biological fixation, the living component of the soil being initially responsible.

Biological Fixation

The many varieties of soil micro-organisms which inhabit the soil require amongst other inorganic elements, phosphorus for the satisfaction of their anabolic requirements. Thus in soils containing a developing microflora, there will be a demand for phosphorus to set against the requirements of the higher plants. Whilst some organisms have been shown to possess the ability to dissolve and use insoluble phosphates, in the main it is considered that the same sources are used by soil bacteria as are used by growing plants. This means that a proportion of the plant available phosphorus in the soil will be diverted and locked up in the living bacterial protoplasm. On the death and decomposition of these organisms phosphorus compounds will be released to the soil, but their immediate or ultimate availability has been little investigated. Even if this phosphate were fully available there would be an overall fixation of phosphorus during periods when the microbiological population of the soil is increasing and this may occur during periods of maximum requirements by crops. However, if this does become freely available in the course of time it will mean a gradual increase in the level of available phosphorus in the soil, allowing for that removed in harvested crops, as the micro-organisms do appear to utilise some at least of the more insoluble forms, which thereby become added to the pool of available nutrients.

Chemical Precipitation

The chemical precipitation of soluble phosphate occurs in soils having a pH sufficiently high for the phosphates of calcium and possibly magnesium to be insoluble. At these pH values, which range between about 6 and 8·5, calcium and magnesium are important exchangeable cations and have a significant ionic activity enabling the precipitation reactions to proceed. As indicated on page 46, it is not easy to define a precise reaction between the alkaline earth cations and orthophosphate ions in the soil but as a general rule the higher the pH the lower will be the solubility of the alkaline earth phosphate combinations.

Iron and aluminium phosphates are notable for their insolubility on the acid side of neutrality, and in acid soils having pH values lower than 6 the precipitation of phosphates in these forms is often claimed to occur. It is in fact improbable that there is a significant concentration of ionic iron or aluminium in any but the most acid of soils, and under moderate acidities the 'free' iron or aluminium is present as an amorphous colloidal hydroxide gel. This being so, a direct precipitation of phosphate as ferric or aluminium phosphate is unlikely to occur.

Anion Exchange

Combinations of phosphate with iron or aluminium do, in spite of the preceding remarks, constitute the biggest proportion of insoluble phosphates in most mineral soils. They are formed by interaction between phosphate ions in solution and the insoluble hydrogels of the iron and aluminium. The phosphate ions displace hydroxide ions from the surface of the solid phase, becoming part of it, the hydroxyl ions appearing in the solution. This process is known as anion exchange and it may also occur at clay mineral surfaces. In the case of the sesquioxides the combination may be expressed as $R(OH)_{3-y}$ $(H_2PO_4)_y$, where y has a value between 0 and 1, R representing either Fe or Al. These phosphate sesquioxide associations probably develop with age and may eventually form the more well defined mineral phosphates as described on page 45.

The nature and degree of exchange occurring on the soil clay minerals is difficult to assess. Laboratory experiments with kaolinite have shown that phosphate ions can be removed from a solution containing them and that the reaction is much less marked with montmorillonite. As in the former there are exposed sheets of hydroxyl ions and in the latter the only surface hydroxyls are those at the crystal edges, an anion exchange reaction seems a reasonable hypothesis. However, it has now been shown that phosphate associations with kaolinite lead to the liberation of silica and it seems likely that under acid conditions kaolinite will tend to decompose slightly with the splitting off from the lattice of aluminium ions. These, partially hydrated, will tend to offset the acidity by becoming exchangeable cations allied to the kaolinite crystal. The presence of phosphate ions will cause the precipitation of this exchangeable aluminium as aluminium phosphate with the re-creation of the acid clay which will further dissociate into the alumina and silica components. Laboratory experiments have shown that kaolinite is 'soluble' in alkali phosphate solutions and that aluminium phosphate is a product of the solution. This means that phosphate fixation by clay minerals may be a form of aluminium precipitation.

(*iii*) POTASSIUM

Alongside nitrogen and phosphorus, potassium is an element of importance in connection with soil fertility because it is required in relatively large quantities by growing plants (see Table 1). Within the plant potassium does not combine to form any specific compounds of compositional importance but apparently exists, associated with organic and inorganic ions, in solution where it assists in many of the metabolic reactions taking place.

Soil mineral matter normally contains a relatively high content of potassium as shown by a full chemical analysis. The amounts vary, of course, according to soil type, siliceous sandy soils having low total potassium contents in the region of 0·1 per cent or less, while clay soils may have upwards of 4 per cent total potassium. In many instances soils having high levels of total potassium support crops which show evidence of potassium starvation while healthy growth can be found in soils which are characteristically low in total potassium. There is no direct relationship between the total potassium content and a soil's power to supply the element in quantities sufficient for plant growth.

Primary Sources of Potassium

Many primary and secondary minerals contain potassium as a constituent element, the more important of them from the point of view of soil studies being the orthoclase and microcline felspars, the muscovite and biotite micas, the clay mineral illite and, of lesser quantitative significance, glauconite. The position of potassium in the lattice structures is described on p. 16 for felspar and muscovite mica. The biotite structure is similar to that of muscovite except that biotite is trioctahedral (see p. 51), having magnesium and ferrous iron substituted for aluminium. The illite structure is also similar to that of muscovite except there is less silicon substitution by magnesium in the tetrahedral positions with a corresponding reduction in potassium to balance the latter charges. It is less well crystallised and to some extent the potassium has been displaced from its theoretical position by other ions. Glauconite is less easily defined but is a silicate mineral analogous to the micas variously containing magnesium, ferrous and ferric iron in the octahedral layers.

These minerals do not give up their potassium in any simple way, and the structural atoms must be regarded as being unavailable from the point of view of plant nutrition. It was seen in Chapter 2 that primary minerals do slowly decompose under the various weathering influences and that during such decomposition alkali metals, where present, may be liberated. Potassium ions are produced by such weathering processes although the rate of weathering does vary according to mineral type. From the point of view of potassium liberation biotite mica decomposes more readily than muscovite mica, which in turn breaks down quicker than orthoclase felspar, microline being the most stable. The greater ease of liberation of potassium from the mica minerals has been explained in terms of crystal structure. In these minerals the potassium ions are situated along the cleavage planes of the mineral where they are easily exposed to surface effects and can be removed by cation exchange processes, whereas in the felspar the potassium is within the crystal lattice, held by covalent bonds, and can only be liberated when the entire structure is disrupted.

The potassium ions produced by the weathering process can suffer one of two fates. On the one hand the potassium can be lost completely from the soil, washed through by naturally percolating waters eventually to drain into the sea. On the other hand it can be retained within the soil and this portion of the potassium is of direct concern in soil fertility studies. This potassium will be absorbed onto the colloidal constituents of the soil, entering into exchange with ions already present and displacing them to be leached from the profile. Such potassium is regarded as 100 per cent available to plants.

Exchangeable Potassium

It is not easy to define quantitatively exchangeable potassium as under natural conditions it appears to exist in a dynamic equilibrium between several states. A natural soil is a phase system involving crystalline minerals, amorphous colloids and a liquid soil solution phase. An intermediate condition between the true liquid phase and the solid phase exists in the form of colloidally bound water and the exchangeable ions are, in the main, associated with this latter medium. Due to ionisation of exchangeable ions and the simple solution of soluble salts the truly liquid phase contains potassium in true solution. Under normal conditions the concentration of potassium in true solution varies between 1 and 10 parts per million and may represent anything from 0·5 to 5 per cent of the total 'easily removable' potassium (which includes the so-called exchangeable potassium). That this soluble potassium includes some ions dissociated from the colloidal soil material is indicated by the fact that the proportion of 'easily removable' potassium appearing in the true solution phase can be increased considerably by aqueous dilution of the soil. The distinction between soluble and exchangeable potassium would appear therefore to be somewhat vague. This partition of potassium depends upon three main factors, which are:

(a) The concentration of potassium in true solution; if it increases the equilibrium moves towards exchangeable potassium.

(b) The nature of the exchange material; humus has a greater affinity for alkaline earth cations and releases potassium more readily to the solution, whereas the mineral colloids have a greater affinity for the alkali.

(c) The nature and relative concentrations of other cations present in the system.

Potassium Fixation

It was shown by DeTurk and his co-workers that if all the easily removable potassium was leached from a soil using, say, dilute hydrochloric acid solution, after the leached sample had stood for several days there was a spontaneous generation of exchangeable potassium within it. This process can be repeated several times and each time there is a regeneration of an exchangeable potassium fraction, the fractions becoming progressively smaller until eventually the process ceases to operate.

The converse of this phenomenon also takes place and if a soil is moistened with a solution of a potassium salt and dried, the subsequent leaching of the soil to remove exchangeable potassium will not immediately lead to a 100 per cent recovery. Some of the potassium is retained in a form not easily displaced by the ordinary extracting solutions, and the conception of a 'difficulty exchangeable' fraction has arisen, the atoms having a role somewhat between the 'structural' and the 'exchangeable'. It is imagined that the atoms involved take up sites around and within the crystal lattices, getting jammed in positions from which they do not readily free themselves. The hexagonal gaps in the basal oxygen sheets of silica tetrahedra would furnish likely holes into which potassium ions would fit tightly. Such fixed atoms are eventually liberated when the concentrations in the outer regions of the system become depleted. Soil potassium relations can perhaps be illustrated simply by the equilibrium system:

$$\text{Structural Potassium} \rightarrow \text{'Fixed' Potassium} \underset{\leftarrow}{\rightarrow} \text{Exchangeable Potassium} \underset{\leftarrow}{\rightarrow} \text{Soluble Potassium}$$

Plant Utilisation of Potassium

It is useful in the appreciation of the status of mineral nutrients in soil to have a picture of the way in which these elements enter into the plant.

The active root hairs responsible for mineral nutrient absorption can be visualised as having a protoplasmic covering, which must be semi-permeable in character, as it permits the through passage of water and preferentially allows certain substances to pass whilst barring free access to other substances.

From the soil solution, the diffusion of cations and simple anions through the root membrane is obviously possible as indicated by the completely successful growth of plants in nutrient solutions. It is, however, improbable that diffusion of this sort is sufficient to account for the full needs of a plant grown in soil, as the concentration, for instance, of potassium ions in soil solution rarely exceeds 20 parts per million, whereas the concentration considered generally appropriate for nutrient solution culture is of the order of 500 parts per million.

Growth in culture solution provides evidence of a cation exchange process associated with the roots. As the base cations are absorbed by the plant the culture solution becomes progressively more acid, indicating the release of hydrogen ions to the solution from the roots. This process must also occur during growth in soil and the hydrogen ions liberated into the soil solution will be able to exchange with the exchangeable cations of the soil and to some extent will help refurnish the soil solution with the cations necessary for plant growth. Even with this mechanism in operation the soil solution is not likely to be able fully to meet crop needs, especially during the active growing season when ion absorption is exceedingly active.

There must be a further mechanism for cation supply to the plant and considerable evidence exists for what is termed 'contact exchange' between soil and plant.

During growth the root hairs must be largely in contact with soil material, and an intimate contact between protoplasmic membrane and soil particles is established. So intimate is the contact that it is often impossible completely to separate the two and the colloidal soil material merges imperceptibly into the root hair surface which is also colloidal in character.

The colloidal root surface carries biologically produced hydrogen and the colloidal soil materials carry basic cations. These cations are

F

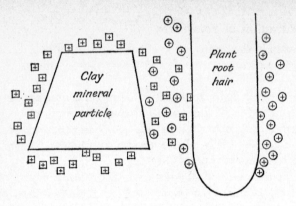

FIG. 13.3 Diagrammatic representation of contact exchange involving cations.

not static, but are in a constant state of motion due to their inherent thermal energies, so that when two such colloidal surfaces approach so closely that the vibrations of the soil cations overlap with the thermal vibrations of the root hydrogen ions, a transfer of parentage, as it were, becomes a matter of no difficulty and the clay and humus cations become readily associated with the plant protoplasm and can be immediately utilised by the plant for any necessary biological processes.

It is considered that this contact exchange process (Fig. 13.3) is responsible for supplying a large, if not the major, proportion of a plant's mineral requirements.

(iv) TRACE ELEMENTS

In addition to nitrogen, phosphorus and potassium there is a long list of other elements that have been shown to be essential for the full and proper development of higher plants. Calcium, magnesium, iron and sulphur are required as structural components of plant tissues but in quantities not so great as the first mentioned elements, and in the normal course of events the soil is adequately supplied with them. Most soils also contain enough of the other elements necessary for growth, the trace elements. These include manganese, boron, molybdenum, copper, zinc and others which are required in minute quantities to act as catalysts and co-enzymes in the metabolic reactions of the growing plant. While these elements may be required in quantities amounting only to a few parts per million of plant tissue, their presence is just as essential to the plant as is the presence of adequate nitrogen, phosphorus and potassium.

Owing to the minute amounts of the trace elements actually absorbed by plants, the average soil is usually amply supplied although, as in the case of the major nutrients, their presence in a soil does not necessarily mean that they are in an available form, as is well illustrated in the case of manganese which is an almost invariable soil constituent but is frequently 'deficient' in certain soils which have alkaline reaction.

In connection with the trace elements, sometimes called micronutrients or even minor elements, it is also important to appreciate that many of them, when present in an available form in quantities significantly in excess of plant requirements, can be harmful to growth or can have secondary effects which are undesirable. Elements such as boron, copper and zinc are actually toxic to plant growth in higher concentrations and other elements such as molybdenum and selenium, while not necessarily damaging growing crops, do cause considerable harm to stock fed upon such crops; excess selenium causes the so-called 'alkali disease' in cattle and the 'teart' pastures in the West Midland Counties are a consequence of high molybdenum absorption by pasture grasses. Other indirect trace element troubles can also arise as, for example, a deficiency of available cobalt which, although it does not affect plant growth adversely, does have harmful effects on certain animals which may have to graze cobalt deficient herbage.

Calcium, Magnesium, Iron and Sulphur

These elements are not trace elements in the accepted sense of the term, although with the exception of calcium they may be considered alongside them. The role of calcium and magnesium in the soil is discussed more fully in the next chapter, and it may be sufficient here to say that calcium is more important as a soil conditioner than as a plant nutrient, and that when the amount of exchangeable calcium, which is also plant-available calcium, falls to a nutritionally low level other growth inhibiting factors are likely to be dominant. Calcium deficiency in plants has been recorded, however, and usually occurs in very sandy soils with a low exchange capacity.

Magnesium deficiency in the field is also associated with light soils in which the exchangeable magnesium has reached a low figure. It more commonly occurs in horticultural practice where the higher plane of fertility normally encountered accentuates any shortage of magnesium.

Although iron constitutes about 5 per cent of the earth's crust and the element is universally distributed, there are certain soils, predominantly calcareous, on which iron deficiencies may occur. In such soils the iron compounds have such a low solubility due to the alkaline environment that some crops, and in particular some tree fruits. cannot absorb sufficient for their needs. In such cases application of soluble iron compounds to the soil is not an efficient antidote because of the rapid precipitation and fixation of the iron which takes place and it must be applied directly to the growing crops in a form which can be absorbed through the growing tissues. The use of certain organic iron compounds for soil treatment has been recommended

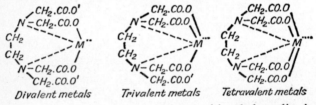

Fig. 13.4   Structures of metal complexes with ethylene-diamine-tetra-acetic acid.

and used with varied success. The compounds used contain chelated iron or iron which is bound to the molecule in an un-ionisable form and the more commonly used are ferric dipotassium ethylene-diamine-tetra-acetate (Fe—EDTA) and related compounds. Fig. 13.4 indicates the structure of Fe—EDTA. Another such compound is the iron complex with di(hydroxy ethyl)ethylene diamine diacetic acid (Fe—HEEDDA). These compounds are absorbable by plants, and the iron they contain, being in an un-ionised state, is not precipitated in the soil medium.

Deficiency of sulphur is rare. Cases have been reported of tea plants suffering from a shortage of sulphur, but under general agricultural conditions this state of affairs is unlikely to occur owing to the fact that sulphur is a major ancillary element in common inorganic fertilisers such as ammonium sulphate and ordinary superphosphate which contains about 50 per cent of calcium sulphate. Soil dressings of organic manures also contain appreciable quantities of sulphur, originating in the sulphur-containing amino-acids of animal and vegetable proteins.

The Trace Elements

One very important characteristic of trace elements is that the total amounts in soils vary widely due to their geochemical distribution. Soil contents of a single element may range from less than 1 part in 10 million to more than 1 part in 100. The major nutrients, on the other hand, never vary much more than, say, fivefold in the amounts found from soil to soil.

However, as with the major nutrients, the level of availability of the trace elements is not necessarily related to total content, although, because of the very small quantities concerned, estimations of the total amounts present in the soil may throw light on the potential supply of available forms of the nutrient. Knowledge of the total amount of element present, used in conjunction with other information such as soil pH, the organic matter content, and the oxidation or reduction potential of the environment, may make it possible to forecast the likelihood of excess or deficiency. The forecasting of trace element conditions may also be assisted by knowledge of the geological background of the soil allied to knowledge concerning the geochemical distribution of trace nutrient elements. Table 16 gives some idea of relative proportions of some elements in the surface layers of the earth.

TABLE 16
GEOCHEMICAL DISTRIBUTION OF ELEMENTS IN THE LITHOSPHERE (after Goldschmidt)

	Percentage by weight		Percentage by weight
Oxygen	46·6	Manganese	0·1
Silicon	27·7	Boron	0·001
Aluminium	8·1	Cobalt	0·004
Iron	5·0	Molybdenum	0·00023
Calcium	3·6	Copper	0·007
Sodium	2·8	Selenium	0·000009
Potassium	2·6	Zinc	0·008
Magnesium	2·1		

In the following paragraphs a selection of trace elements will be discussed, to illustrate the general background of their existence in the soil and the factors influencing their availability to plants.

Manganese. Of the trace elements manganese is most abundant, soil contents ranging from around 1 part per 10,000 to 1 part per 100. The manganese content of plants is also very variable, being generally between 50 and 500 parts per million of dry matter. Although individual plant species show a range of tolerance, if the manganese content falls below about 5 parts per million of dry matter, growth restriction due to manganese deficiency can be expected; whereas if the content of manganese rises to levels greater than 1,000 parts per million of dry matter, growth limitation due to manganese toxicity is likely. There is no relationship between the total manganese content of the soil and the plant.

The manganese absorbed by plants is taken up as the divalent manganese ion and divalent manganese may exist in soil as an exchangeable cation or as carbonate, bicarbonate or oxide. The element also exists in forms exhibiting valencies of 3, 4, 6 and 7 and the trivalent and tetravalent manganese oxides Mn_2O_3 and MnO_2 occur in many soils. Manganese dioxide is very insoluble and offers no available element while the availability of Mn_2O_3 appears, from *in vitro* studies, to depend upon its physical state, the amorphous form being useful, although the ageing of the precipitated compound gives reduced availability, probably due to the formation of a more stable crystalline form. It would seem that the higher oxidative states have a low availability and that the manganese-supplying power of a soil depends upon the proportion of its manganese which is in the divalent state.

This proportion is to a large extent controlled by the general soil environment, being influenced by pH, degree of aeration and biological status. The 2, 3 and 4 valency states appear to co-exist in equilibrium, the end states of which are the divalent and the tetravalent. At higher pH levels (greater than 7), and when a predominantly aerobic flora dominates the soil, the equilibrium moves towards the tetravalent form. Acid soils and anaerobic environment favour the stability of divalent manganese.

Boron. Tourmaline is the principal primary mineral containing boron and it is widely distributed throughout the earth's crust, occurring in many soils as a heavy mineral in the sand and silt fractions. It is a very stable mineral, resistant to natural weathering, and can make no direct contribution to the boron needs of growing plants. The range of extractable boron in soils ranges from about 1 to 100 parts per million and probably this originates in the slow weathering of tourmalines and allied minerals. This available boron exists as borate and differs from the majority of trace elements in being anionic rather than cationic. As an anion, borate is not retained in the soil by any physico-chemical reactions and its persistence must depend upon the formation of sparingly soluble salts which must be susceptible to biological breakdown. Borates of calcium, magnesium or other metals may be the immediate sources of available boron, but in a calcium saturated soil with a pH between 7 and 8·2 the solubility of the boron compounds becomes much reduced and a deficiency of the trace element may be induced by overliming.

Soils derived from shales, loess and alluvium are usually high in available boron, and soils from a sandstone parent material are often intrinsically low in the trace element and are more susceptible to an induced deficiency.

Molybdenum. Molybdenum is another element which is presumed to be absorbed by plants in anionic form as molybdate. The element occurs in soils as an ancillary element in olivine, in clay minerals, and in a few other accessory minerals. It is thought to be slowly released by natural weathering when it is most likely to take an oxidised form as a molybdate. The normal level of extractable molybdenum ranges from less than 0·1 to about 5 parts per million, although in certain areas, due to particular geological formations as exemplified by the Lias clays in Somerset, the level may rise to around 200 parts per million.

Plant requirements for molybdenum are very small, and the uptake of the nutrient is very much influenced by soil reaction, its availability decreasing markedly as acidity increases. The liming of a soil can counteract a deficiency where absolute shortage of the element is not the limiting factor. Maximum uptake of molybdenum by plants appears to take place between pH 7 and pH 8.

Cobalt. The trace element cobalt has not been shown to be essential for plant growth and development, but its importance is associated with the essential role it plays in animal nutrition. It appears, for instance, that herbage grazed by sheep should contain cobalt contents not lower than 0·08 parts per million of dry matter in order to avoid the possibility of pine disease in sheep and cattle. Cobalt is normally associated with basic igneous rocks and soils derived from these or their sedimentary deposits are rarely deficient. Acidic igneous rocks and in particular certain granites and their derived sandstones can be inherently lacking in cobalt. In the soil itself little is known of the cobalt status, although the element appears to be associated with both the organic and inorganic colloidal fractions.

Zinc and copper. These metals occur in soils principally as chance lattice elements in the crystal structure of many minerals, including the clay minerals. On liberation from the lattices as they disintegrate during weathering, they are retained as exchangeable cations in the exchange complex of the soil or linked to the organic matter in some chelated combination. Nutritional deficiencies of these elements seem to occur mainly on peat soils, but they have been reported on mineral soils. They appear to be mainly limitations of availability, but little precise information is available concerning the mechanisms and reactions involved.

14   Soil acidity

Soil acidity has already been mentioned in connection with certain soil formation processes and in connection with the availability to plants of phosphorus compounds in the soil. It has, however, other far-reaching effects on plant growth, most of which depress the potential yields of economically valuable crops, and it has been estimated that, in this country, more crop failures, either complete or partial, are due to soil acidity than to any other single cause.

Some soils are naturally alkaline in reaction; immature soils for instance, developed from chalk or limestone parent material, or soils developed under climatic conditions where evaporation exceeds precipitation with a consequent surface accumulation of alkali and alkaline earth salts. The majority of soils, however, tend to have a reaction on the acid side of neutrality, due in the main to the continuous leaching of bases by the percolation of rainwater.

pH

The acidity/alkalinity status of a soil is referred to as the soil reaction, and as all reactions in the soil take place in an aqueous environment, acid/base reactions must be examined in this context. Soil reaction is therefore summarised by reference to a suspension of soil in water or occasionally in a dilute solution of a neutral salt such as calcium chloride.

Dissociation of water. Pure water, as is well known, dissociates into two ions, the hydronium ion $(H_3O)^+$ which is associated with acidity, and the hydroxyl $(OH)^1$ which is connected with alkalinity.

$$H_2O + H_2O \rightarrow (H_3O)^+ + (OH)^1$$

In aqueous media the free acid cation is invariably present as the hydrated proton $(H_3O)^+$ but for convenience it is often referred to as a hydrogen ion and written H^+, and the dissociation of water then becomes

$$H_2O \rightleftharpoons H^+ + OH^-$$

This is a reversible reaction and the law of mass action applies so that the concentration of hydrogen ion $[H^+]$ multiplied by the concentration of hydroxyl ion $[OH^1]$ divided by the concentration of undissociated water $[H_2O]$ is constant. Because the degree of dissociation of water is so small, the concentration of undissociated water molecules always stays essentially the same and can be considered as being constant also, so that $[H^+] \times [OH^1]$ equals a value (Kw) known as the dissociation constant of water. This can be measured, and for pure water at 21° C has a value of 10^{-14} using concentrations measured in terms of gram-equivalents per litre. As $[H^+]$ and $[OH^1]$ are equal in the dissociation of pure water the values of each concentration must be 10^{-7} mEq/litre.

If substances are added to or dissolved in water, which gives protons to the system or absorbs them, then the concentration balance between H^+ and OH^1 is disturbed. The numerical product of the concentrations must, however, remain constant so that if $[H^+]$ increases, $[OH^1]$ must decrease accordingly and vice versa.

To simplify the arithmetical representations of this important equilibrium Sörensen suggested that the state could be defined simply by reference to the hydrogen ion concentration, and to avoid the use of cumbersome decimal fractions he proposed the use of a value designated pH (Puissance d'hydrogen) which he defined as the $\log_{10}$ of the reciprocal of the hydrogen ion concentration measured in grams per litre.

$$pH = \log_{10}\frac{1}{(H^+)} \quad \text{or} \quad -\log_{10}(H^+)$$

In these terms the pH value of water equals 7. When the pH value drops below seven the solution becomes more acidic and when greater than seven it becomes more alkaline.

Soil Acidity

With the possible exception of very acid organic soils, soil acidity is centred around the colloidal fraction (the clay and the humus constituents) in the soil material. The exceptional cases have a certain amount of free sulphuric acid present due to action of oxidative bacteria on sulphur compounds pre-existing in the soil as iron pyrites or organic sulphur compounds.

While the clay and humus colloids are both active seats of soil acidity, the clay contribution is the one that has been most extensively studied, and the following account refers more specifically to mineral colloids, although the general principles involved pertain to the organic colloids also.

It was seen in Chapter 4 during part of the discussion on clay minerals that associated with each colloidal particle was a complement of cations there referred to as exchangeable cations. These cations could be all of one kind or could be mixed according to circumstances, and it is with the cations that soil reaction is directly concerned. The individual colloidal particle was shown to be negatively charged and the exchangeable cations had the function of neutralising the negative charges of this colloidal micelle. While some of the cations would take up positions within the crystal structure, the majority would be external to it, associating themselves with the outer surface to establish the most stable arrangement possible.

If such a colloidal particle is immersed in water, each cation associated with it will possess an ionisation energy acting to separate the cation from the negatively charged micelle and to give the cation

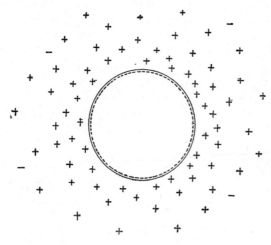

FIG. 14.1   Diffuse-layer system.

a free existence within the medium. Free existence in the solution will, however, be limited by the relatively large negative charge of the particle which will tend to attract cations, and the resulting position of these ions will be determined by the relative strengths of the two forces pulling in opposite directions. The precise position of individual ions, however, will not be constant due to the thermal energy oscillations, and at any particular moment of time some ions will be nearer the surface than the mean position and some will be further away and will have a relatively freer existence to exhibit their ionic properties. Fig 14.1 gives a diagrammatic representation of the diffuse-layer system thus formed which is often referred to as a Gouy system after L. Gouy who endeavoured to define the system mathematically.

In the case where all the exchangeable cations were hydrogen
ions, an aqueous suspension of clay would be expected to behave as a
weak acid, as indeed it does in many ways, the weakly dissociating
hydrogen ions conferring the acid properties. When the exchangeable

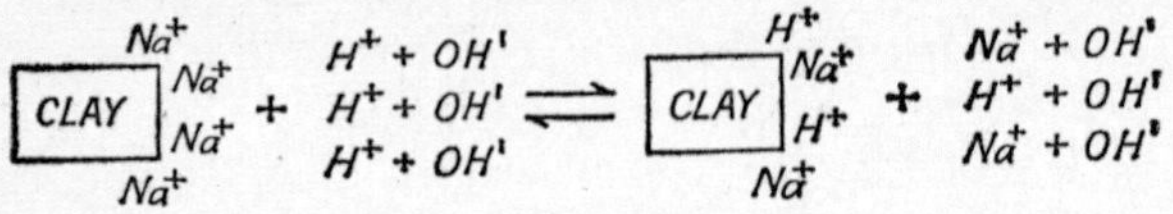

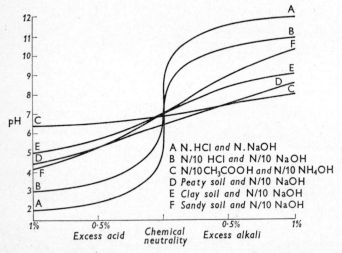

FIG. 14.2   Hydrolysis of sodium-clay.

cations are alkali cations, e.g. sodium, the reaction of the clay suspension is alkaline which is again explicable if the hydrogen-clay system
acts as a weak acid. In this case the sodium-clay ionises partially, the
clay anion forming the un-ionised hydrogen complex with hydrogen
ions from the weakly ionised water, thus leaving an excess of hydroxyl
ions in the suspension to confer alkaline properties (Fig. 14.2).

A   N. HCl and N. NaOH
B   N/10 HCl and N/10 NaOH
C   N/10 CH$_3$COOH and N/10 NH$_4$OH
D   Peaty soil and N/10 NaOH
E   Clay soil and N/10 NaOH
F   Sandy soil and N/10 NaOH

FIG. 14.3   Titration curves.

That the properties of hydrogen-clay and acid peat systems are
similar to but not absolutely identical with weak acids is shown by
comparisons of titration curves of acid soils when the pH is plotted
against successive additions of alkali. Figure 14.3 gives an example
of the type of comparison obtained. The curves in the figure show
how the clay and humus are markedly buffered to changes in pH,

whereas the strong acid (hydrochloric acid) and to a lesser extent the sandy soil offer little resistance to pH change after successive increments of alkali have been added.

The difference between a clay acid and acetic acid is noticed in the shapes of the two curves they produce. Whereas the acetic-acid curve is smooth, the clay-acid curve has distinct changes of shape, indicating different states of association of the clay-acid hydrogen ions and reflecting the different types of charge of the clay crystals as discussed on page 54. The buffering properties of the soil colloids are very important from the point of view of plant development and for the maintenance of a reasonably stable bacterial flora, both of which are very susceptible to changes in soil reaction.

Considering the broad picture of the soil composition it will be evident that the pH of a soil suspension, as measured by hydrogen-ion concentration, will not be uniform as it would be, for instance, in a solution of dilute hydrochloric acid. In the soil suspension as in the soil itself there will be regions around the clay and humus particles where ionic concentrations are high, and regions between particles where the concentrations of ions are much less.

Many different techniques for measuring soil pH have been suggested, including the use of colorimetric indicators and electronic pH meters. Many variable factors influence the results obtained by all methods and it is unwise to quote results to a higher accuracy than 0·1 pH unit. Most field soils are tested using indicators whilst in the laboratory suspensions of soil in water are taken for electronic pH measurement. The dilution of the suspension does not have a major effect on the result as soils are fairly well buffered and the difference between 1 : 1 and 1 : 10 soil water mixtures is unlikely to be more than 0·2 pH units, being slightly higher in the more dilute suspension. The presence or absence of carbon dioxide will give greater variations. In order to increase the reproducibility of pH measurements however, a standard practice is recommended which suggests that a 1 : $2\frac{1}{2}$ soil water suspension, allowed to stand overnight to equilibrate with the atmosphere, should be used.

Variations in measurements can be minimised by using a dilute solution of calcium chloride instead of water as the suspension medium. This, a neutral salt, compresses the double electric layer and reduces the concentration gradients of the ions around the colloidal particles. Measurements made by this technique are significantly lower than those of aqueous suspensions and although more reproducible as laboratory measurements are not as popular as the more traditional techniques when it comes to interpreting soil fertility problems.

The pH measured in the field or in the laboratory is an average value but within the soil itself there will be appreciable variations in pH between points which are really quite close together. This is actually of great significance in permitting a wide range of different types of micro-organisms to co-exist in the soil, and may influence local fixation or liberation of available plant nutrients, and it may also explain why certain plants survive long after the overall soil reaction has changed from a favourable to an unfavourable state.

Factors Influencing Soil Reaction

Soil reaction is influenced by three considerations involving the soil colloids and their associated exchangeable cations. These are:

(a) The nature of the acidoid (a name given to the pseudo-acid radicals in the colloidal make-up of the soil).

(b) The ratio between H ions and basic cations in the total content of exchangeable cations.

(c) The relative proportions of the different basic cations present.

(a) **The nature of the acidoid.** The hydrogen-saturated clays of different clay-mineral structures exhibit differences in pH. Montmorillonitic clays, for instance, usually exhibit lower pH values than kaolinitic clays. This is due to the difference in origin of the bulk of the particle change; in the montmorillonitic clays the changes arise largely from isomorphous replacements within the lattice, whereas in the kaolinite group of minerals the charges originate almost exclusively at the crystal edges and are centred largely at edge oxygen atoms. Hydrogen ions compensating at charged oxygen sites dissociate less easily than those held to offset internal charge discrepancy, and therefore the activity of the hydrogen ions is restricted in the 1 : 1 lattice types which, in consequence have higher pH values (less acid) than the 2 : 1 lattice clay minerals.

Humus acids give rise to lower pH values than mineral colloids, due to the apparent greater strengths or dissociations of these acids.

The clay fractions of many tropical soils frequently do not show high acidities even in regions where the leaching of the soil is very intensive. This is due to the fact that the clay fractions of such soils are not of the silicate mineral types as previously described. They are largely colloidal ferric and aluminium hydroxides which show little acidity due to the almost negligible dissociation of the exchangeable hydrogen ions in their constitution.

(*b*) **The ratio between hydrogen and basic cations.** It has been shown that a hydrogen-saturated soil is acidic and an alkali-saturated soil reacts alkaline due to a hydrolysis reaction such as is associated generally with salts of strong bases and weak acids. When the exchangeable cations of the colloidal complex are mixed the resultant reaction will be the compromise of acidic and alkaline reactions. In soils developing under humid, temperate climates the elements found as the principal basic cations are calcium, magnesium, sodium and potassium and a very average ratio of these cations would be of the order Ca : Mg : Na : K : : 80 : 10 : 7 : 3. This means that the basic cations are dominantly calcium ions and any reaction which touches upon the basic ions of the soil will reflect most upon the calcium status.

Speaking again in the most general terms, a soil will react neutral when the ratio of hydrogen ions to basic cations in the total exchange capacity is of the order 40 : 60.

It was pointed out in Chapter 8 that under natural conditions there is a tendency for bases to be leached from the soil and replaced by hydrogen ions with a consequent increase in soil acidity. It is found that mineral soils do not naturally develop especially low pH values and figures less than pH 4 are rarely encountered. If mineral soils are artifically acidified to a lower degree than this they revert in a short time to their natural minimum. The explanation of this is that the crystal lattice of the clay minerals is unstable in very acid media, decomposing with the liberation of silica and alumina. The lattice aluminium atoms thus freed are capable of replacing exchangeable hydrogen ions to become exchangeable aluminium, the pH thereby being increased by this incorporation of basic ions to a level at which the lattice is once more stable.

(*c*) **The proportions of different basic ions.** The character of the bases involved in the exchange complex influences soil pH quite appreciably, due to the different degrees of dissociation of their complexes with clay colloids. Sodium clays, for instance, hydrolyse much more strongly than calcium clays and thus give a more alkaline reaction. Calcium soils reach a maximum pH in the region of 8·2, whereas in certain alkali soils where sodium is the chief cation the pH may attain a value of the order of 10. It is therefore apparent that the proportions of alkali and alkaline earth cations will be important in determining the pH level of a soil. In temperate regions such as the

British Isles, and in colder climates calcium ions are the most important and dominate the scene.

Soil Acidity

Success of plant growth is, amongst other things, intimately associated with soil reaction, and the optimum reaction for the majority of agricultural crops is in the range pH 6·0 to 6·5. A number of crop plants will grow successfully in soils more acid than this, but in the main the maintenance of a soil reaction around pH 6·5 should be the farmer's aim.

The adverse influence of soil acidity on plant growth is due to a variety of causes, any of which may assume greater importance in any particular case. As acidity increases, exchangeable calcium decreases, and a plant deficiency of calcium may ensue. The increase in hydrogen ions may cause a direct toxicity effect. More important, however, are the indirect effects: the reduction of available nitrogen as nitrifying organisms are repressed; the reduction of available phosphate following fixation with soluble iron and aluminium; the direct toxicity of aluminium ions as they become significantly present in the exchange complex. There are effects too on the trace elements of the soil, and in particular the increase in soluble divalent manganese induced by greater acidity may prove toxic to some plant species, and a reduction in the solubility of molybdenum has been shown to occur, with the possible onset of molybdenum deficiency to reduce certain crop yields.

Liming

Soil acidity thus brings about a complex of undesirable features. Rectification of this acidity is usually brought about by the incorporation of basic calcium compounds which may be either calcium oxide, calcium hydroxide or calcium carbonate. Occasionally magnesium compounds are used and these may be desirable when magnesium deficiencies exist.

There seems little to choose in effectiveness between the various liming materials, calcium carbonate (provided that it is in a sufficiently fine state of division) being as effective as its chemical equivalent of calcium oxide. Dolomitic limestone appears just as effective in reducing the effects of high soil acidity as carboniferous limestone, although magnesium oxide in significant dressings can bring about a sterilising effect on the soil due to its caustic alkalinity. Table 17 gives some experimental results showing yield responses of sugar beet to chemically equivalent dressings of different liming materials; Table 18 compares increasing dressings of calcium oxide and a magnesian lime.

The correction of soil acidity is important, but there are dangers

TABLE 17
THE INFLUENCE OF CHEMICALLY EQUIVALENT DRESSINGS OF DIFFERENT LIMING MATERIALS ON SUGAR BEET YIELDS
(kilogrammes per hectare)

District	No Lime	Carbon- iferous Lime (CaO)	Magnesian Lime (25% MgO)	Carbon- iferous Limestone CaCO$_3$	Magnesian Limestone (30% MgCO$_3$)
Ravenstone	0·30	8·76	11·70	9·04	7·38
Ercall Heath	0·00	13·06	17·07	12·58	15·72
Hatfield	5·65	7·86	10·00		

associated with excessive use of lime when, in many instances, it is possible to raise the pH of the soil to figures of the order of pH 8. At this level of pH plant growth troubles may be induced by reducing the availability of certain nutrients, amongst which the trace elements

TABLE 18
INFLUENCE OF INCREASING DRESSINGS OF CALCIUM OXIDE AND MAGNESIAN LIME (25% MgO) ON YIELDS OF SUGAR BEET
(kilogramme per hectare)

Dressing	Nil	1	2	4
Carboniferous Lime	5·65	8·01	7·71	9·17
Magnesian Lime	5·65	9·84	10·22	7·33

manganese and boron in particular are significant. It becomes important therefore to obtain some assessment of the amount of lime necessary just to neutralise the excess acidity, without incurring an uneconomic wastage of liming material with the potential dangers associated with excessive dressings of lime.

The measurement of 'lime requirement' presents a problem that has been studied extensively, and various techniques have been suggested and are variously used, although the problem of lime requirement determination is not so straightforward as might be expected.

The detection of soil acidity is a simple task. The appearance of a dark-coloured solution when soil is shaken with dilute ammonia (Schutz's test), or the production of a blood red colour when shaken with an alcoholic solution of potassium thiocyanate (Comber's test) both indicate acidity. An actual measurement of pH gives a better idea of the intensity of acidity and this can be done on a soil water suspension, using an electrometric pH apparatus, or more approximmately by means of indicators.

That such tests in themselves give no information as to the amount of lime necessary to apply in order to correct the pH to an optimum of 6·5 should be apparent, although the pH measurement is often assumed to be correlated with lime requirement. In certain instances a correlation can be established, but, as it has been shown, the pH is related to the proportion of hydrogen ions to basic cations in the exchange complex and not solely to the absolute quantity of hydrogen ions present. This means that two soils may have the same pH value and yet have widely different requirements of lime to bring about neutralisation. A consideration of two hypothetical examples will best illustrate this principle. If soil A has a cation exchange capacity of 100 milli-equivalents per 100 grammes of soil, of which 60 milli-equivalents are hydrogen ions and 40 are calcium, it will have a pH on the acid side of neutrality. If soil B has an exchange capacity of only 10 mEq cations per 100 grammes of soil and the proportion of hydrogen to calcium is 6 mEq hydrogen to 4mEq calcium, then B will again be acid and its pH value will be similar to that of soil A. If each soil is brought to neutrality, the proportions of hydrogen to calcium will have to be adjusted to approximately 40 : 60 and soil A will require 20 mEq of hydrogen to be replaced by calcium, whereas soil B will only require 2 mEq of cation exchange. Thus for these soils having similar pH values, because of their different constitutions, the lime requirement of one is about ten times that of the second. Fig. 14.4 demonstrates graphically the lime requirement differences associated with different soil textures.

Quantitative determination of lime requirement cannot therefore be based simply on pH and several alternative techniques are available.

The amount of exchangeable calcium present in a soil has frequently been used as a guide to liming both alone and in conjunction
with cation exchange-capacity measurements. For soils of known type
the exchangeable calcium figure alone may be a reliable guide as for
example certain soils of North Wales where liming is considered
advisable when the figure falls below 0·03 per cent. It would be
misguided, however, to use such a figure to cover all cases and
Hissink in Holland proposed a more general scheme, to measure the

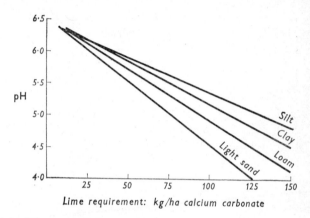

FIG. 14.4   Lime requirement curves.

degree of calcium saturation or the 'V value'. The relationship
$$V = \frac{S}{T} \times 100,$$ where S is the exchangeable calcium and T the exchange capacity, was suggested with the general proposal that a lime
requirement was indicated when V dropped to a value lower than fifty.

A further method of approach is to try and measure quantitatively
the acidity by means of a neutralisation technique. Direct alkali
titration in the customary manner has obvious difficulties. Addition
of a neutral salt solution to an acid soil will lead to an exchange
between the salt cation and the hydrogen ions of the soil with the
formation of an acid solution which, after filtration to remove the
soil material, can be directly titrated with alkali. This 'exchange
acidity' can be then correlated with lime requirement. The exchange
acidity is, however, the product of a reversible reaction and therefore
will be influenced quantitatively by the relative amounts of soil
and neutral salt concentration, and the extent of cation retention by

the soil will vary according to the kind of cations involved (see page 166). Thus methods of this sort are very arbitrary in character, the results being difficult to reproduce accurately, and they depend for their utility on the correlation of experimental observations in the field with very strictly defined laboratory conditions.

A more commonly used variant of this theme is the measurement of the alkali neutralised by the soil, after an excess has been added. The most commonly used solution is calcium bicarbonate in a method suggested by Hutchinson and McLennon. Calcium bicarbonate solution of known strength is mixed with soil and after the reaction has taken place the strength of the residual bicarbonate is measured and hence the calcium absorbed by the soil.

$$\boxed{\text{soil}} \begin{matrix} H^+ \\ H^+ \end{matrix} + Ca\,(HCO_3)_2 \rightarrow \boxed{\text{soil}}\ Ca^{++} + 2H_2O + 2CO_2$$

The method is not free from disadvantages; controlled experimental conditions in regard to strength of reagent, proportions of reactants and time of reaction must be defined, or variable results will be obtained. As the pH of saturated calcium bicarbonate solution is of the order of 6·2 it is only possible to estimate the lime requirement to bring the soil to that pH, and thus in some instances the method may not be applicable, although for general agricultural purposes it has proved very useful in the past, giving results that can readily be related to field practice.

The construction of a soil titration curve is also quite useful for determining lime requirement. In this technique known amounts of soil are mixed with increasing amounts of lime in solution and the corresponding pH values plotted on a graph. It is then possible to calculate the amount of lime necessary to adjust the pH of a sample to any desired figure. Owing to the differing buffer capacities of soils, each sample should be treated individually but the method has certain advantages. It is not necessary to take a large number of readings as the significant portion of the titration curve involved is the approximately linear portion between pH 4 and pH 7 (see Fig. 14.3).

With all methods of lime requirement determination the results have to be multiplied by a 'field factor' which can only be found by linking laboratory determinations with field trial observations, and in most cases this value lies between 1·5 and 2.

15   The deterioration of soil

The determination of the soil's potential for useful utilisation by man is a serious problem. The agriculturist has, over the centuries, developed techniques whereby the productive capacity of virgin soil has been increased beyond all measure, and this high productivity is now a *sine qua non* for man's survival. In increasing soil productivity and by bringing into cultivation more and more land, an artificial situation has arisen whereby the high level of soil fertility is maintained by a complicated series of treatments and managements which are in calculated opposition to natural tendencies. Any errors in the treatment or the management of the soil can therefore quickly lead to a destruction of fertility, which unless quickly recognised and countered can give rise to much human distress.

Soil deterioration falls under two broad headings which can conveniently be discussed separately although they are not infrequently complementary to each other. The first involves loss of fertility in the generally accepted sense, whilst the second involves the actual loss of fertile soil and is more familiarly known as soil erosion.

Deterioration of Fertility

The most important factor in fertility deterioration involves the natural leaching of the soil and the consequent development of acidity, as was discussed in more detail in Chapter 14. The lime losses of the soil become pronounced in regions of higher rainfall and in the vicinity of industrial smoke, and unless measures are taken to counteract the increasing acidities, low fertility will ensue. Under natural conditions the flora and fauna are able to adapt themselves to developed acidity and an equilibrium is established leading to a stable natural economy. If husbandry practices are carried out on acid land, the deterioration may be enhanced by destruction of soil structure and the intimate humus mineral association will be disturbed. In addition, the potentiality of the soil for immobilising plant nutrients will be increased, leading to serious economic loss.

Such a situation can be remedied by the appropriate use of lime, fertilisers and manure and the use of proper cultivations and cropping.

A further form of deterioration concerns the natural water regime of the soil. This may be affected in two ways, i.e. by water becoming excessive or by the water supplies to the crop becoming restricted.

Disturbance of drainage may occur by the production of an

indurated layer in the subsoil regions causing a perched water table to develop which brings about a waterlogging of the soil volume containing plant roots with a reduction of growth. Indurated layers may form naturally as 'iron pans' in soils undergoing podsolisation or the obstructive horizon may be brought about as a 'clay pan' which not infrequently forms in heavier soils at plough depth. The use of appropriate subsoil ploughs to break such obstructions is the obvious reaction to these situations when they are recognised. Occasionally drainage difficulties are induced by the continued use of heavy implements on clay soils when there is a possibility of padding down the soil so that structural elements tend to form with a horizontal cleavage, effacing the vertical cracks normally found in such soils which contribute so much to their drainage. No simple remedy to such cases has been found as yet.

Deterioration of fertility often follows the injudicious use of water-soluble fertilisers and the constant and overzealous application of materials containing sodium, as, for instance, nitrate of soda or certain grades of potash salts, may help destroy the ability of clay soils to form tilths suitable for seed beds. This is due to the fact that the sodium ions, if applied in excessive amounts, enter and may become dominant in the exchange systems, and sodium clays easily lose structure to form a deflocculated mass which dries out with the formation of wellnigh indestructible clods. The problem is similar in a smaller degree to that occurring when land is damaged by invasion of sea-water or when areas of land are to be reclaimed from the sea. Initially, when such areas become free from excess water and begin to dry out they can be cultivated and form admirable soils from the point of view of structure in spite of having a cation exchange complex dominated by sodium. This is to be accounted for by the presence of a high concentration of free sodium chloride in the soil solution which is sufficient to keep the sodium clays in a flocculated form. As this surplus sodium chloride diminishes, washed out of the soil by rainwater, the structures of these soils collapse and they become unusable agriculturally until the exchangeable sodium is largely replaced by calcium.

To effect an early replacement of sodium by calcium is the object of restoration schemes, but calcium carbonate is not the most appropriate source of calcium for this aim as the displaced sodium would furnish sodium carbonate by the double decomposition type of reaction and this, being highly alkaline, would adversely affect plant growth and might also affect the humus status of the soil. Calcium sulphate in the form of gypsum is the ameliorant most commonly used, being sufficiently soluble to provide a steady source of calcium ions for displacing sodium. The sodium sulphate formed, being a neutral salt, will not affect the soil during its transient existence therein.

In dry and arid regions, where irrigation is practised, and not infrequently in the horticultural industry soluble salts accumulate in the surface soils, there being insufficient drainage to remove salts surplus to plant requirements. As the concentration of these salts increases, so the osmotic pressure of the soil solution increases until it interferes with the water transpiration through the semi-permeable plant root membranes and plant survival is jeopardised. Improvement of water relationship is the only remedy. Associated with this condition of high salinity is the probable establishment of an imperfect nutrient balance which also brings about a reduction in a soil's production potential.

The direct determination of soluble salts present in a soil can be made, but it is technically inconvenient and the assessment is usually done indirectly, use being made of the increase in the electrical conductivity of water brought about by the presence of ions in solution. The indirect method is arbitrary, the electrical conductivity of an aqueous suspension of soil being measured and expressed in terms of specific conductivity.

It is usual to measure the specific conductivity in terms of the reciprocal ohm or mho per centimetre, but as the mho is an inconveniently large unit for soil salinity work the sub-unit millimho is more often used. The determination is frequently made on a 1 : 5 soil water extract and at a defined temperature as solution conductivities change about 2 per cent for each degree centigrade variation from a reference temperature of $25°$ C.

As conductivity figures are often a little cumbersome the use of a pC notation has been suggested in analogy to pH and pF. The value pC is defined as the common co-logarithm of the specific conductivity of soil extracts or suspensions measured in terms of mhos/cm, and has the virtue of giving simple values for comparative purposes.

Practical interpretations can also be adequately made by translating conductivities into direct measurements of soluble salts either as percentage total salts or as milli-equivalents of salt per cent. The correlation between salt content and conductivity of soil extract can only be approximate as considerable variations in the make up of soluble salts appear from soil to soil. Nevertheless approximate relations exist, and for practical purposes the specific conductivity (mhos/cm) at 20°C multiplied by 375 gives the percentage total soluble salts.

The toxic effects of soluble salts in plant growth become noticeable within the range 0·1% to 0·2%, and the boundary between 'saline' and 'non saline' soils is usually taken at 0·1%.

Soil Erosion

The erosion of land surfaces, whereby loosely consolidated fragments of the earth's surface are moved by natural agencies and ultimately deposited on the sea-bed, is a phenomenon that has existed since the earth became a solid mass, and is ceaseless in its operation. It is a part of the natural order of things. Under normal conditions the rate of erosion is slow and its effects pass largely unnoticed, except that open drains become silted up and, especially noticeable after heavy rain, rivers and streams can be seen as turbid suspensions of eroded land surfaces.

There are occasions, however, when this erosion of the land becomes catastrophic in character and the living soil is removed too quickly for its usefulness as a growing medium to be exploited for the benefit of man. When this happens to husbanded land it becomes a very serious matter. Apart from cataclysmal events over which no control can be exercised it is possible for an acceleration of the erosion process to be brought about by ill-advised use of land and by introducing factors into the natural equilibrium which disturb it so as to emphasise the erosive element. In the past large areas have been disturbed in this way, as for example, by injudicious deforestation and removal of the protective cover afforded by trees without any attempts being made to substitute an alternative protection; in many regions, too, fertile areas of land have been intensively cultivated with the effect that the humus content of the soil has been subjected to loss by destructive oxidation and the structure has collapsed to render the soil much more susceptible to movement under the influence of erosion forces.

Water and wind are the two main factors which lead to these soil losses.

Water Erosion

The ability of rain to bring about surface erosion depends upon two things. First, the dispersal of the surface particles by the beating action of the raindrops; second, the power of the surface run-off water to transport the dispersed material. Of these two things the latter has the more positive effect in actually carrying away the surface soil. If there was no run-off there would be no erosion.

Run-off is again the consequence of a combination of circumstances. The factors involved are the ability of the soil to absorb the precipitation falling upon it and its ability to drain the water away when it has been absorbed. Investigators in this field of study recognise an 'infiltration capacity', which is the rate at which water can pass into the soil, and 'percolation capacity', which is the rate at which water passes through the soil, and these two rates are not necessarily the same. In a dry soil the infiltration capacity is usually much higher than the percolation capacity, but as the soil becomes more and more saturated with water the two values become more nearly equal until the stage is reached when the rate of water infiltration is more or less controlled by the rate of percolation. This means that precipitation falling on a wet soil gives a higher run-off than does the same precipitation falling on a dry soil.

It is also true that run-off is directly related to draindrop size. Fine rain infilters more easily than does rain falling in large drops which can reach a size around 6 mm in diameter. This is in large measure due to the actual physical action of the raindrops falling on to a soil surface. Large drops mechanically disperse the soil crumbs by reason of their beating action, giving a dispersed suspension of soil, the fine particles of which lodge in the surface cracks and fissures, effectively blocking the soil pores and forming a relatively impervious crust to the soil and bringing about an increased run-off. This breakdown of the crumb structure also has the effect of rendering the soil material much more susceptible to movement in the slowly moving water streams on the surface.

A further consideration of the erosive action of rainfall indicates that an intensity factor, rather than the total precipitation, dominates the extent of possile erosion, the more intense the rain the greater the erosion. This is obviously connected directly with disperson and run-off, and is very well illustrated by some figures obtained in 1935 by Hayes and Palmer of the Soil Conservation Experiment Station, Wisconsin, and referred to by Baver (Table 19).

As would also be expected, topography has a marked effect on the extent of soil erosion. On perfectly level land surfaces erosion would be expected to be slight, if any occurred at all, whereas on a sloping surface the run-off would be facilitated. As the speed of water flow varies directly with the degree of slope, and as the carrying power of water increases in proportion to as much as the fifth or sixth power of its velocity, according to detailed conditions, even a moderate slope can help bring about heavy losses of surface soil.

Erosion on a sloping surface is also influenced to a certain extent by the force of gravity. There can be simple slipping of the material of the surface which has been loosened by various means, or movement may be caused when, as during heavy rainstorms, splashing of the raindrops occurs. The rebound of raindrops from a surface is practically at right angles to the surface, and on the rebound, the drops are charged with fine soil material which is carried upwards with the splash. On falling back this solid material falls to a point further down the slope than it originated.

TABLE 19
RELATIONSHIP BETWEEN SOIL EROSION
AND INTENSITY OF RAINFALL

Duration of Rainfall	Amount of Rain	Loss of Soil calculated as tonnes/hectare
30 h 35 min	6·5 cm	1
1 h 52 min	4·75 cm	130

The denudation of soil surfaces by water as just described falls into three classes for purposes of qualitative descriptions. The classes are: sheet erosion, rill erosion, and gulley erosion.

Sheet erosion is usually the most insidious form of erosion from the agricultural point of view, as it proceeds largely unnoticed. It is prevalent on only slightly sloping country and takes place by the removal of a small but approximately uniform layer of land surface. During some heavy storm the entire soil surface becomes dispersed and the muddy surface water is collected and led from the site via surface drains. There is no significant change of contour and the losses take place insidiously until, maybe, the whole top soil is removed before any appreciation of danger is felt.

Rill erosion is a further stage in the intensity scale of erosion. It is accompanied by the formation of small channels or rills which direct the surface run-off to more formal drainage channels. Whilst the rills cut their way through the surface soil they do not become so deep as to escape obliteration during subsequent land cultivation.

Gulley erosion is, in a sense, an exaggerated rill erosion. It occurs when channelled run-off water has acquired sufficient body and velocity to gouge a channel which may be anything from 1 to 30 metres in depth and 6 to 300 metres in width. The gulleys may be formed after cloud-bursts and normally dry off after the rains, forming dry channels. They are initiated most often on steeper slopes during torrential downpours of rain and subsequently form natural drainage channels for surface water, becoming thereby spontaneously enlarged.

Protective Measures against Water Erosion

The principles upon which protective measures to combat water erosion are based are designed to prevent soil disperson during wet periods and to reduce surface run-off. The dispersion of the soil is minimised by the development of a granular structure with stable aggregates. Lime and organic matter are important in this connection. Vegetative cover to break the fall of dropping rain and reduce the effectiveness of its dispersive force whilst at the same time giving root systems to bind together the soil mass is again important. Grass is perhaps more generally effective in this connection, being operative all the year round. Arable cropping has to be so arranged that effective cover is provided, especially during the periods of greatest rainfall when erosion forces are at their highest.

The reduction of run-off can variously be effected according to circumstances. Terracing of cultivated land in horizontal steps, contour ploughing to make the plough ridges completely horizontal, giving no slope for rills to develop, contour drilling of cereals and row crops assist in like fashion to impound the water until it naturally drains through the profile. Strip cropping is often practised, whereby crops giving effective soil cover are grown alongside crops not so good at protecting the soil so that overall erosion control is achieved.

Wind Erosion

Wind erosion can be very devastating where it seriously occurs, and dust storms removing large volumes of soil can sometimes completely obscure daylight. The wind-blown soil is often deposited on other land surfaces, leading to deposits of loess or dunes in littoral regions. Wind erosion most frequently affects soils with a particulate structure and fine sands are very susceptible. Certain organic soils with a mor type of humus which has partially oxidised and had its cohesive properties reduced, often become 'fluffy' and tend to blow.

Increasing the colloidal content of such soils often reduces the propensity to move under the influence of wind force and farmyard manure or clay marl applications may often stabilise the surface by assisting the formation of crumbs. Table 20 gives the mechanical analysis (American scale) of four soils in the Vale of York, illustrating how soils with slightly higher clay contents are immune from blowing.

TABLE 20

RELATIONSHIP OF SUSCEPTIBILITY TO WIND EROSION
AND MECHANICAL ANALYSIS

	Soils suffering wind erosion annually		Soils not normally blown	
	a	b	c	d
Coarse sand	54·1	44·2	31·8	35·0
Fine sand	33·8	47·3	57·4	40·4
Silt	1·2	0·6	1·8	3·0
Fine silt	1·0	0·7	0·7	2·3
Clay	2·6	3·1	4·4	6·2

A more general measure against wind erosion is the planting of belts of trees to form wind breaks. Milder forms of wind erosion may have little effect on well-established crops, but may tend to remove germinating seedlings, and in such circumstances seedlings can often be protected by sowing them between ridges ploughed at right angles to the direction of the prevailing wind.

16   Soil analysis and field experimentation

The soil is a natural object, differing in form and properties where it exists under different conditions, and capable of changing its form and properties should the conditions of its environment change. It is also capable of being modified artificially to increase its usefulness to mankind. It is the material on which grow the land plants of the earth needed to feed and nourish the varied populations it supports.

Attempts to analyse the soil chemically and physically can be made with one or two objects in view. It can be analysed to define its objective properties which can then be used in an endeavour to identify its kind; mechanical analysis and mineralogical analysis are examples of such determination. In the second case, analyses to determine its ability to support food or other necessary crops can be attempted, and more energies are usually devoted to this latter end, not necessarily because of the economic issues involved but often because of the greater difficulties encountered.

The complex nature of soil fertility has already been discussed, and it is usual to restrict subjective laboratory analyses to the determination of the reaction of the soil and determinations of the levels of available nutrients. Problems of water-supplying ability and the influence of structure on growth have not yet proved to be amenable to general laboratory solution.

Even in the apparently more simple exercises of determining reaction and available nutrients there are many difficulties. Some of these are connected with the lack of precise knowledge of what actually defines the availability of plant nutrients, and lead to the acceptance of analyses based on arbitrary soil extractions which give figures which require 'interpretation' before they can be usefully used.

The question usually asked of the scientific adviser is whether or not the application of some treatment will give increased returns which will be economically justifiable. The only basic and fundamental way to obtain the answer to such a question is to experiment on a small scale under field conditions using plots of land to simulate the field scale.

Single Plot Experiments

In earlier days when attempts were made to get exact information about the results of soil treatment the obvious thing was done. Two areas were marked out; one received the treatment in question, and the other did not, and the quality and yield of the crops from the two areas were compared. Many of these experiments have confirmed the value of traditional practice and many have initiated new techniques now in common use. There are some of these experiments whose results are evident beyond dispute; there are others in which the results have been doubtful, varying from place to place and from year to year. When, for example, the yield on the treated plot is constantly something of the order of 50 per cent greater than that from the untreated plot and still more when the difference is visible to the eye and the line of demarcation between the plots is seen in the crop, there is no doubt that the treatment has effected an increase. On the other hand, when the differences are small, say 5–10 per cent, and no visible demarcation between the plots is seen in the crop, no reliable conclusions can be drawn. Such small differences may very well arise

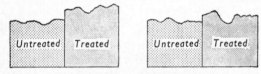

FIG. 16.1   Assessment of treatment differences.

even if the plots are not treated differently, for the lack of uniformity in the crop-producing power of practically all soil areas is considerable and much greater than is apparent to the eye. It appears to be doubtful whether even experienced agriculturists can detect a difference of 15 per cent or even 20 per cent between two adjacent crops. Choosing uniform land for crop experiments is therefore not as easy as might appear. Possibly the uniformity of ripening of a crop is the best criterion.

If any particular soil treatment has a positive effect on a crop there will in the ideal case be a sudden increase in quantity or some point of quality as one passes from the untreated to the treated plot. The determination of that sudden change is the crucial problem in deciding whether the treatment has or has not an effect. Care should be taken clearly to distinguish between the two extreme cases illustrated diagrammatically in Fig. 16.1; in one of which an abrupt change is obvious and in the other of which that change is obscured by its own smallness and by soil variation influences. The case in which the abrupt change is obvious is straightforward qualitatively and is not usually difficult to deal with quantitatively. The case in which the abrupt change is small and obscure can only be dealt with—even qualitatively—as a probability determined by replication in various ways.

Testing the Uniformity of the Land

Attempts have been made to overcome this difficulty by treating the two plots similarly for a year or two before initiating the experiment. Thus on the Manor Farm at Garforth in Yorkshire there was a series of meadow hay plots, differently manured, and it was decided to examine the additional effect of lime to be applied to half the area of each plot. For three years before the liming the yields of hay from the two halves of each plot were recorded and the average difference between them was allowed for in estimating the increase due to liming. Some of the initial differences between the two halves were as high as 1,250 or 1,500 kg per hectare.

This preliminary testing of the ground is very useful, but unless it can be carried on for a number of years it may not help as much as might be thought, for considerable differences in the differences are sometimes found from year to year.

Replication of Plots

The replication of plots in various places and the continuance of them for several years may admit of conclusions being drawn where that would be impossible from any one of the experiments in any one year, and it is usual in laying down experiments designed to test any treatment which is not likely to give an overwhelming difference to replicate the plots. This replication of the plots, however, introduces further difficulties, for the treatment which is being investigated may have a definite and demonstrable effect in one place and may have no such effect in another. For example, in the many trials of various phosphates and slags which have been carried out in this country there is abundant evidence that the phosphate has affected the crop in a marked manner at some centres and not at others. The duplication of the experiment *in another place* does not necessarily help towards the decision as to the effect of the treatment at the original centre. With sufficient data from properly replicated experiments the statistician may be able to state the probability of a particular treatment having an effect on some untreated land. That, however, is a different and less useful thing than being able to differentiate between those areas in which the treatment is likely to yield positive results and those areas in which it is not. Such information, and at best only limited, can only be expected from the study of the results of the replicated experiments taken in conjunction with a study of the type of soil involved.

The Half Drill Method

A method of replicating plots on one field and which is mainly of use in comparing two varieties of corn is illustrated in Fig. 16.2. In variety trials the seed of one variety is put in one half of the drill box and the seed of the other in the other half. While this method involves

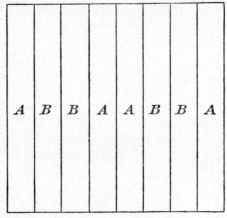

FIG. 16.2   The Half-drill arrangement of plots.

some practical harvesting difficulties it has great advantages over a mere replication of larger plots. It is possible, for instance, to form some idea of soil variations by noting any regular increases or decreases along the series of plots of the same variety all similarly treated.

The Interpolated Standard

A method of arranging plots is illustrated in Fig. 16.3 which is useful when there is evidence of a regular variation of the soil at right angles to the plots. The plots A_1, A_2, etc., are replicates of the 'control' plot—i.e. the untreated plot or the standard variety. The plots a, b, c, etc., represent different treatments or varieties. If there is a regular soil variation in the direction of the arrow, the yield of a control plot where 'a' is would be $A_1 + \dfrac{A_2 - A_1}{3}$ and the ratio of the yield of 'A' with that of a control plot on the same ground is

$$A_1 + \frac{a}{\dfrac{A_2 - A_1}{3}} \quad \text{or} \quad \frac{3a}{2A_1 + A_2}$$

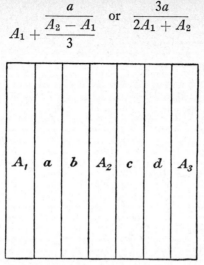

FIG. 16.3   The 'Interpolated Standard' method of plot experiment.

Balanced Plot Method

Figure 16.4 illustrates an arrangement of plots in triplicate. The feature of the arrangement is that the distance in one direction of one

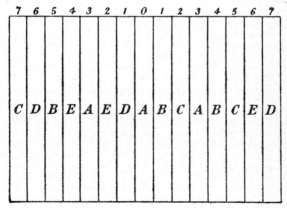

FIG. 16.4   The 'Balanced Row' arrangement of plots.

of three similar plots from the middle is equal to the sum of the distances of the other two in the other direction.

G

Latin Square Technique

The methods described above cater for those cases where there is a systematic change in soil conditions occurring in a definite direction. In the majority of fields significant changes occur sporadically in random areas and these are more difficult to assess. A device introduced many years ago attempts to deal with such situations. It is illustrated in Fig. 16.5, where four different treatments or varieties are catered for, but the square can be adapted to accommodate

A	B	C	D
D	A	B	C
C	D	A	B
B	C	D	A

Fig. 16.5   The 'Latin Square' arrangement of plots.

more or less. The four plots are arranged in quadruplicate in such a way that each occurs once in each vertical series and once in each horizontal series. The sums of the yields from the four different treatments may be obtained in eight different ways, by four vertical and four horizontal additions. If the land were perfectly uniform and the experiment entirely devoid of errors, these eight totals would be identical. In practice they will vary and the application of simple mathematical considerations to the extent and manner of the variation admits of some conclusions being drawn about the variation in the land and a correction being made to the observed yield in each of the sixteen squares. The method fell into disfavour among agriculturists on account of harvesting difficulties but it is now used quite frequently and gives very useful results.

The squares can be arranged in a large number of ways to fulfil the requirement that no one treatment appears more than once either in the vertical or the horizontal series. To avoid any suspicion of bias in the arrangement the design used in any one instance is obtained by a random choice from the full range of possible arrangements.

Randomised Block Method

When the number of treatments to be tried becomes large, or when it is desirable for such treatments to be replicated several times, the Latin Square Method becomes very cumbersome, and the Randomised Block arrangement may be used.

In this method the experimental area is divided into blocks of equal size and equal in number to the number of replications to be used. Each block is then subdivided according to the number of treatments under trial. The actual position of a particular treatment within a block is chosen at random.

FIG. 16.6   Randomised block method.

For instance, suppose five different treatments are under trial and each is to be replicated eight times, i.e. 40 plots in all. The experimental area is divided into eight compact blocks. Within each block one plot will be assigned at random to each of the five treatments. If the treatments are lettered A–E a possible lay-out would be as shown in Fig. 16.6.

The advantage of the method is that a high degree of replication can be obtained and it is easy to distinguish variations between blocks due to soil variation from those within blocks due to different treatment. On uniform land the totals of the various blocks will be comparable as each block contains one plot of each treatment. Differences between the block totals can be ascribed to soil heterogeneity and these variations can be eliminated from the variations due to treatment.

The proper control of field experiments in which relatively small differences of yield are found is a much more intricate matter than was formerly supposed, and in designing an experiment of any importance it is advisable that a statistician be consulted to ensure that the true significance of the results can readily be ascertained and that the experimental design is such that errors incidental to all such work can be assessed accurately.

Pot Experiments

The difficulties attendant upon the lack of uniformity of soil conditions may be very largely overcome by carrying out experiments with small quantities of soil in suitable pots after previously mixing thoroughly all the soil involved in the experiment. Pot experiments, however, are not so simple as is frequently supposed. While the soil used may be made uniform by artificial mixing, the difficulty of filling several pots so that the degree of compactness will be the same in all is very great. It is not usual to obtain any satisfactory conclusions from pot experiments in which small differences are involved unless the experiment is carried out at least in quadruplicate. Moreover, the conditions under which the soil is held are artificial, and the important connection with other soil round it and with the subsoil and rock below is broken.

Useful results can, however, be obtained by carefully conducted pot experiments, but they frequently involve more trouble than many field experiments and it is not always possible to transfer the results obtained to practical circumstances.

The Laboratory Examination of Soils

The use of field experimentation to solve problems of soil behaviour is expensive and time consuming and it is not surprising that the agricultural scientist is often asked to advise on the basis of laboratory experience and analysis.

The first problem to be overcome in this endeavour is the difficulty of obtaining a suitable sample of the field or fields in question that will be representative of their composition. Unless this can be done, laboratory studies can have no practical value to the farmer. The technique of field sampling is thus of prime importance. In an investigation into field sampling errors Hemingway found that 24 spot samples in an apparently uniform field ranged in values for citric acid soluble phosphate from 5 parts per million to 265 parts per million. The available potassium figures of the same samples ranged from 105 to 800 parts per million and the pH values from 5·46 to 7·10.

To get a fair picture of the soil status therefore demands the examination of a large number of spot samples so that a reliable average composition can be achieved. To prevent laboratory work getting out of hand it is customary to bulk the spot samples collected to give one average sample upon which analysis can be made following appropriate mixing and pretreatment.

The detail of collecting the spot samples must vary with each field to be examined but a few general principles must be observed. The number of spot samples taken should not be too small, 10–15 from

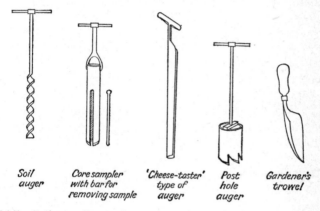

| Soil auger | Core sampler with bar for removing sample | 'Cheese-taster' type of auger | Post hole auger | Gardener's trowel |

FIG. 16.7   Soil sampling tools.

each sampled area being appropriate for areas not exceeding 6 hectares. If fields are larger than 6 hectares it is better to subdivide them, taking bulked samples from each smaller area. If there are any obvious variations in a field the appropriate areas should be separately sampled; areas of marked topographical difference within the field or areas differentially cropped, fertilised or fenced off should be noted in this connection. Notes or explanatory sketches of such divisions should be made at the time of sampling. Areas or irregular shape should have the spot samples taken with a random distribution except in so far as spots likely obviously to bias the bulk sample should be excluded. These include footpaths, regions in the immediate vicinity of gates, sites of recent manure heaps etc. Uniformly shaped areas are best sampled along a zig-zag line across the direction of the cultivations or fertiliser drills.

The actual spot samples can most easily be taken with a screw type auger (Fig. 16.7) but a cheese borer, a small trowel or a post-hole type auger can be used. The screw type auger is screwed into the soil to the requisite depth, pulled directly out and the soil trapped in the thread of the auger removed as the individual sample. On loose dry soils the small trowel may prove less frustrating. The depth of sample to be taken varies with the purpose. For nutrient status of arable soils 15 cm to 25 cm depths are usual, whereas on grassland the upper 7·5 cm may be more appropriate. On permanent grassland to be broken up or for other special reasons two or more samples are often taken separately, say 0 to 7·5 cm and 7·5 to 15 cm.

At the time the samples are taken it is often of great help in the subsequent interpretation of the results of laboratory tests to obtain further general information concerning the field. Information as to the crop rotation and the previous cropping is important; notes on topographical features which might indicate liability to frost pockets or flooding, wind exposure, etc., are of value; comments on the system of drainage, its efficiency and the condition of ditches, symptoms of nutritional troubles, signs of plant pests and diseases, notes of the common weeds and nature of grass swards should all be made when conditions permit.

In the laboratory the bulk samples are air dried and carefully crushed to separate soil aggregates without breaking stones which are removed from the samples by means of a 2 mm sieve (see page 25). The fine earth is then carefully mixed and is ready for analysis.

Sufficient has been said earlier to show that interpretation of the results of soil analysis is difficult and somewhat obscure. Nevertheless it is possible to obtain information in the laboratory as to the possible response of a soil to a particular treatment.

Mechanical analysis considered in conjunction with the lime status, humus content and water conditions is of value in considering treatment to meet difficulties of texture and tilth. It sometimes enables the agriculturist to say whether the difficulties in question are inherent in the soil itself or arise from external influences such as situation and climate. Again, mechanical composition is an important factor in deciding whether heavy land shall be laid down to grass or whether liming and careful management may admit of its being retained under arable cultivation or in the case of very light sandy soils their suitability for crops such as carrots or the susceptibility to wind erosion.

The chemical analysis of soil is of more value in some cases than others. The usefulness of lime requirement methods in deciding whether the lime status of the soil is adequate for any proposed crop has been discussed (p. 171). The value of determining the so-called 'available' phosphorus and potassium by extraction with acidic extractants is perhaps less, but it is nevertheless appreciable.

Laboratory methods of examining soils are still empirical and the application of their results to agricultural problems is still somewhat tentative, although with the increase in data available, and its correlation with field experiments, by the increased areas of soils which have been surveyed and mapped, very useful progress has been made.

The Use of the Plant in Determining Manurial Requirements

When it was first realised that the chemical analysis of soil gave results that could not be interpreted in any simple way, some attention was given to the possibility of making use of the analysis of plants grown in the soils in question. An endeavour was made to ascertain whether the percentage of phosphorus and potassium in the ash of the plant could be used as a guide to the manurial requirements of the soil in which it had grown. There was little advance on these lines until Neubauer suggested a method which has proved quite workable in practice.

Neubauer's Seedling Method

In this method the soil is mixed with pure sand, and rye seedlings are grown for 17 days from date of sowing. On the eighteenth day the shoots are removed and the potassium and phosphorus in the ash determined. It is claimed that unless these percentages reach values equivalent to the order of 0·025 per cent K_2O and 0·006 per cent P_2O_5 there is an insufficiency of available potassium or phosphorus in the soil to produce a maximum crop.

Mitscherlich's Pot Culture Method

Another method of this type is based upon the view held by Mitscherlich on the basis of experimental pot work, that the amount of any given manurial constituent which is required to produce a maximum crop is independent of the amounts of the other constituents present. For example, if a soil contains adequate amounts of phosphorus and nitrogen but is poor in potassium, increasing application of potassium will produce increased growth up to a certain maximum. If the amounts of phosphorus or nitrogen had been inadequate increasing applications of potassium would still have produced increases in crop growth up to a certain maximum. That maximum would be less than when the phosphorus and nitrogen supplies were adequate, but the amount of potassium required to produce it would, according to Mitscherlich, be the same. On this assumption it is possible by pot experiments in which there are four units, one with an adequate supply of nitrogen, phosphorus and potassium and three others in which each of these manurial constituents is omitted in turn, to compute the manurial requirements of a soil.

The validity of both the Neubauer and Mitscherlich methods is the subject of very considerable discussion.

Aspergillus Niger Method

Another biological method used for determining nutrient status and especially certain trace elements is based upon the growth of a fungal mycelium of Aspergillus niger. A definite quantity of soil is added to a nutrient solution complete in all nutrient elements save the one under investigation. The amount of that nutrient in the soil sample then becomes the limiting factor in the growth of the mould, and after incubation under standard conditions the mycelial development gives an index of the nutrient status of the soil as far as the missing element is concerned.

17   The mapping of soils

In the previous chapter it was indicated that there exists a great diversity of soils and that in order to determine how any individual soil will react to a particular treatment some rather complicated experiments have to be performed. More general application of such experimental results might perhaps be possible if individual soils were defined and maps were drawn indicating where similar soils existed. This task is, however, not easy. The classification systems for soils as described in Chapter 9 are of little value in a detailed study of soil properties and potentials as they only divide soils into groups based on very general characters which are predominantly genetic. Within each sub-group of the broader classifications are innumerable individual soils differing one from another in many ways significant to their capacities to grow crops, and although the aim of soil mapping is not necessarily so utilitarian in its ambition, its practical usefulness will be judged eventually on its success in correlating soils of recognisable productive potential.

There are in existence maps indicating land utilisation upon which inferences as to soil properties may be based but land utilisation is the resultant of soil fertility and economic circumstances and is of little permanent value as an index of soil kind. Geological maps have occasionally been used as a basis for recording soil types on the assumption that the soil is a reflection of the rock from which it was derived. The first published survey of English soils in 1911 covering the counties of Kent, Surrey and Sussex was based on the geological origin of the soils and a fair relationship was achieved. Unfortunately this type of successful mapping can only be realised when the complication of surface drift deposits is absent or when the pedological weathering of the soil horizons is not so advanced as to obscure the influence of parent materials.

In this country the nature of the mineral parent material is recognised as playing a part in determining the nature of the soil but it is the chemical and/or mechanical composition rather than position in the geological succession that is important. The parent materials recognised as being significantly different are listed in Table 21.

If the soil were a body defined by well developed and measurable qualities, the actual mapping could be based upon the representation of the regional distribution of one or more of these qualities. Unfortunately it is not cartographically possible to combine on to one single map all the qualities necessary to enable a full description of an object as complicated as a soil and a further problem has to be faced. To single out for mapping certain properties on the grounds of expediency is not a satisfactory solution however valuable such maps

TABLE 21
SOIL PARENT MATERIALS GROUPED ACCORDING TO THEIR LITHOLOGICAL CHARACTERS

1. Acid crystalline rocks	12. Loess
2. Intermediate crystalline rocks	13. Calcareous sands and sandstones
3. Basic crystalline rocks	14. Siliceous sands and sandstones
4. Ultra-basic crystalline rocks	15. Felspathic sands and sandstones
5. Basic tuffaceous shales	16. Glauconitic sands and sandstones
6. Schists	17. Ferruginous sands and sandstones
7. Slates and hard shales	18. Grey limestones
8. Clay shales	19. Cream limestones
9. Calcareous clays	20. Chalk
10. Non-calcareous clays	21. Peat
11. Silt clays	22. Mixed drifts

may be for *ad hoc* purposes. Soil acidity maps, textural maps, maps of soil phosphate status, etc., may have a use but they are not maps in the true sense of recording distributions of recognisable soils, which, it must be remembered, have to be considered in full profile to be true reflections of the subject.

The Soil Series

At the present time the unit of classification used for mapping purposes is the *soil series*, a unit developed originally in the American Soil Survey. This basic unit is defined as being a group of soils formed under similar conditions from similar parent materials and having similar colours, textures, structures and drainage conditions. Thus the profile as a whole is considered rather than a restricted section of it. The soil series may be subdivided if necessary into *soil types* according to textural variations of the surface soil.

The soil series can only be described by reference to a complete profile and it is necessary to prepare a pit or cutting which presents a complete section from surface soil to parent material before any detailed examination is possible. In the mapping of a particular area the first stage is to establish the different series likely to be encountered. This is best done on the basis of a superficial examination of the area, viewing the general topography and the geological evidence to decide on places likely to be characteristic of larger areas. Natural cuttings and sections found in ditches etc. may be of use in the selection of appropriate sites. At the chosen spot a pit is dug to expose the profile which is described in a systematic manner. The principal series having been established in this way the whole area is then systematically examined, using a screw auger to give the surveyor samples of soil at depth, and by this means the areas of the individual series and the soil types within each series can be ascertained. Rarely is there a sharp division between series, and their boundaries are often somewhat conjectural, being decided upon the basis of considered compromise.

Profile Description

The description of a soil profile should be based on an objective assessment of all the tangible features it possesses and an orderly presentation of data is necessary if confusion is to be avoided. In the first instance the profile should be divided into horizons on the basis of visible characteristics and the nature of the horizon should be indicated by the appropriate symbol. The symbols have been described in Chapter 8. Where several horizons of one kind exist in a profile they are indicated by qualifying the letter symbol by numbers reading from the surface downward. The thickness of each horizon present is to be recorded in sequence and the clarity of boundaries indicated.

Horizon Description

The horizons having been listed, it is necessary for each to be described in some detail and the colour is usually the first property to be recorded. As the colour of a soil horizon varies with its moisture content and can be influenced by the nature and incidence of the light in which it is viewed, standard conditions for colour recording are important. It is usual to record the colours of moist and dry soil using accepted colour names which are defined by reference to Munsell Soil Colour Charts. The Munsell reference colours are based on the spectral components of the colour, its intensity and the shade or variation from neutral; an example of a horizon colour record might read: 0–15 cm A_1 horizon. Greyish-brown (9 YR 5/1 dry: 9 YR 4/3 moist).

The texture of the horizon, which defines the relative amounts of variously sized mineral particles present, is also to be registered in the series description. In the field the texture has to be assessed on the basis of a personal judgement and a textural group assigned. The textural grades generally used are described by symbols which take the form of capital letters to simplify the appropriate group, followed where necessary by a qualifying small letter indicating degrees within a group. In Britain the textural symbols used are:

	Light	Medium	Heavy
Sandy soils	Sa	S	Sb
Loamy soils	La	L	Lb
Silty soils	Za	Zm	Zb
Clay soils	C	Cs	Sc
Chalk soils	K	K	K
Peat soils	.	Pt	

As the texture refers specifically to mineral particles less than 2 mm diameter an indication of stoniness is also necessary to bring in to consideration the larger mineral fragments, and if these have any special distribution within the horizon this should be described.

The nature of the aggregation of particles within the horizon and their structural features are horizon characteristics to be recorded, as also is the physical nature of the soil, its compaction, friability or induration.

Form and disposition of organic matter within the horizon are often important features, and often of greater significance still, is an indication of the drainage status which may be excessive, perfect, imperfect or impeded, and which has many repercussions on the overall horizon characteristics.

An example of a full soil series description will best indicate the principal points to be considered and the following example is taken with this in view.

HOOK SERIES

Profile Description

Locality: Hook, Hampshire. Grid Ref. 41/511049.
Topography: Flat, low lying
Altitude: 17·5 m O.D.
Parent Material Group: Loam
Parent Material: Brick-earth
Major Soil Group: Brown earth. Sub-group: Low base status with gleyed B or C horizon
Land Use: Arable
Profile: Drainage: Imperfect
  0–20 cm Grey-brown (7·5 YR 3/2 moist; 7/5 YR 5/4 dry), fine

sandy loam with occasional rust spots. Slightly stony. Structureless. Worms present.

20–40 cm Brown (5 YR 4/4 moist; 7·5 YR 5/6 dry), fine sandy loam, slightly stony. Many pores. Worms present.

40–78 cm Pale brown (5 YR 4/4 moist; 7·5 YR 5/6 dry), fine sandy loam with occasional grey spots. Slightly stony. Many pores. Worms present.

78–90 cm Pale yellowish brown (5 YR 4/4 moist; 7·5 YR 4/4 dry), fine sandy loam with slight rust and orange mottle. Slightly stony. Many pores. More compact.

>90 cm Brown flinty gravel with brown, fine sandy clay loam. Very compact.

The profile description is supported by analytical information as and when necessary.

The Soil Map

In the first instance the soil series defined as described above is given a place-name from the locality where it was first recorded or where it occurs in a well developed form. If it is subsequently found that similar series have been described by different workers in separated districts and given different names the nomenclature is revised, usually in favour of the first described profile. In this way a collection of series within the country is established and can form the basis of a map and in order to restrict any overlap of series careful comparison of new series descriptions with established individuals has to be made.

Because of the fact that many series often recur at intervals in small isolated areas and are to some extent related by some common factor to neighbouring series it is sometimes possible to simplify the problem of mapping larger numbers of small and different areas by grouping series together on the basis of the common factor.

In regularly undulating districts it is possible for several soil series to appear in regular sequence in rhythm with the topography, and where this occurs the name *soil catena*, first suggested by Milne on the basis of certain tropical repetitions, has been applied to the unit group. Soil surveys usually recognise the existence of series groups varying regularly and in pattern according to variation of moisture status as might occur down a hill slope. A unit of such a pattern is referred to as a *soil association* and is mappable as such and if further detail is required the association can be broken down to the series equivalent on the basis of drainage characters. The field map is based in Britain

upon the Ordnance Survey of Great Britain map drawn to the scale
1 : 63360. In the final map the coloured areas representing individual
series are designed to bring to the fore the major soil groups, e.g.
podsolised soils are represented by red colours, brown earths by
shades of brown, and blue colours indicate gley soils. Organic soils
are usually purple in colour.

The Soil Survey of England and Wales, together with the Soil
Survey of Scotland, publish from time to time soil maps of different
areas which have been surveyed together with detailed descriptive
memoirs of the soils represented.

18   Soil science literature

The general student who pursues a course of study in the science of the soil should acquire three things from that course. First, a general knowledge and understanding of soil science as it is viewed at the time of his course of study; second, a competence to acquire more detailed knowledge of any special part of the subject that may for any reason become his special—even though temporary—concern; third, a competence to follow intelligently the future development of the subject. He must therefore know something of the literature. Naturally the specialist will have a more detailed knowledge of the literature of his speciality, but the more general student should 'know his way about' this literature. This chapter is an effort to direct him, although not perhaps to guide him.

There are many books, journals and abstracts pertaining to the subject other than those mentioned below. The selection of those named is not intended necessarily to imply a meritorious distinction between them and the others, and it is hoped that the selection will not be deemed invidious. The books and journals named are those commonly used, and the latest editions available should be consulted, and the student who is familiar with them will inevitably find reference—particularly in the abstracts—to others.

Books

BAVER, L. D.   *Soil Physics*, Wiley, N.Y. This full account of the physical properties of soil contains many useful references.

BEAR, F. E. *Chemistry of the Soil*, Reinhold, N.Y. A collection of monographs on a number of special topics dealt with at an advanced level. The subjects include soil development, ion exchange reactions, soil organic matter, trace elements, plant nutrients and others.

BLACK, C. A. *Soil Plant Relationships*, Wiley, N.Y. This details the properties of soil which have a particular bearing on plant growth.

BRIDGES, E. M. *World Soils*, Cambridge University Press. A well illustrated account of the characteristic soils of the world and their environmental associations.

BROWNE, C. E. *A Source Book of Agricultural Chemistry*, Chronica Botanica Co., New York. The historical development of the subject from the days of the ancient Greek and Roman philosophers to the days of Liebig and Lawes is described in fascinating detail.

BUCKMAN, H. O. and BRADY, N. C. *The Nature and Properties of*

Soils, Macmillan. A very good general text book on the soil and those properties of soil which influence its ability to support plant growth.

CLARKE, G. R. (with BECKETT, P.) *The Study of Soil in the Field*, Clarendon Press, Oxford. A book of much value to those principally interested in those objective properties of soils which enable them to be characterised and classified.

DUCHANFOUR, P. *Précis de Pédologie*, Masson et Cie, Paris. An excellent presentation of ideas relating to soil forming factors and processes. The dynamics of pedology are especially well illustrated.

GRIM, R. E. *Clay Mineralogy*, McGraw Hill, N.Y. A specialist text dealing with the characteristics and properties of the secondary clay minerals.

HESSE, P. R. *A Textbook of Soil Chemical Analysis*, Murray, London. A most valuable text book, giving a critical appraisal of a wide range of laboratory methods of soil analysis. The discussions relating to the theory and backgrounds of the methods described are exceptionally valuable.

KONONORA, M. *Soil Organic Matter*, Pergamon Press. Currently the most authoritative treatise dealing with the chemistry of soil organic matter.

METSON, A. J. *Methods of Chemical Analysis for Soil Survey Samples*, New Zealand DSIR. This book should be in every laboratory concerned with the chemical analysis of soil giving, as it does, a comprehensive range of techniques to cover all the main requirements likely to be met.

ROSE, C. W. *Agricultural Physics*, Pergamon Press. This is a student's text book dealing with the physics of the environment and covering many aspects of soil physics. The mathematical treatment has been developed with the requirements of the non-mathematician in mind.

RUSSELL, E. W. *Soil Conditions and Plant Growth*, Longmans, London. This may fairly be described as a reference library for the general student and comprehensive necessity for the specialist. It gives a full account, with abundant references, of the science of soil conditions. The historical introduction is a valuable feature.

Journals, Reports, etc.

The Journal of Soil Science (published by the Clarendon Press, Oxford) was first published in March 1949. This journal now appears twice a year and contains original papers dealing with all aspects of research into the properties of soil. The standard of the papers printed is very high and every serious student of the soil should be acquainted with the journal.

Soil Science (a monthly publication of the Williams and Wilkins Company, Baltimore, Maryland, U.S.A.). This monthly journal originated in 1916 for the collection together of papers dealing with soil chemistry, soil physics and soil biology and it has maintained this policy for over fifty years. It is an important source of new work connected with the soil.

Soil Science Society of America, Proceedings is the journal of the Soil Science Society of America, Madison, Wisconsin, U.S.A. This important publication, in addition to reporting the domestic affairs of the Soil Science Society of America, contains a wide range of scientific papers on soil topics. The papers are divided into six divisions comprising: (1) 'Soil Physics', (2) 'Soil Chemistry', (3) 'Soil Microbiology', (4) 'Soil Fertility', (5) 'Soil Classification', and (6) 'Soil Conservation'. It also, as does *Soil Science*, contains critical reviews of new books and publications relating to the soil.

Advances in Agronomy, prepared under the auspices of the American Society of Agronomy by Academic Press Inc., is an annual publication containing collected monographs or reviews on particular topics which are varied from year to year. Many of these monographs relate to soil subjects and present a detailed appraisal of information concerning them as it is appreciated at the time of writing. Many references to original papers are listed.

The Journal of Agricultural Science (Cambridge University Press). Originally appearing in 1905, this quarterly journal was an endeavour to collect together technical and scientific information concerning the science of agriculture. Its terms of reference include the whole of scientific agriculture and it includes many papers and accounts of work done with soil alongside work in animal nutrition and crop protection.

Annual Reports of Rothamsted Experimental Station, Harpenden, Herts. These reports contain current accounts of the work being carried out at this world-famous institution and are useful in keeping their readers in touch with modern trends of thought in scientific studies of the soil.

Plant and Soil (Martinus Nijhoff, Holland) is an international journal published in Holland, dealing particularly with problems involving soil fertility in its wider aspects.

Soil Science and Plant Analysis (published by Marcel Dekker, Inc., New York) presents papers on important developments in soil science and crop production of interest to agronomists, horticulturalists and foresters as well as soil scientists.

Geoderma, an international journal of soil science published by Elsevier Publishing Co., Amsterdam.

The International Society of Soil Science, with headquarters at the Royal Tropical Institute, 63, Mauritskade, Amsterdam, is a frontierless organisation of soil scientists, and meets for learned discussions at intervals in different parts of the world. The deliberations of its meetings are published as *Congress Transactions* at appropriate times, each congress being divided into 'Commissions' of more specialist concern.

There are also various publications of the Food and Agriculture Organisation of the United Nations Organisation, with its headquarters in Rome, which treat on scientific aspects of the soil as they affect the work of establishing backward countries of the world and the creation of higher standards of living.

Many national organisations throughout the world publish regular journals which deal either in part or totally with soil science.

The Annals of the Royal Agricultural College of Sweden, Zeitschrift für Pflanzenern ährung, Düngung und Bodenkunde, Pedology (Russia) and *Soil Science and Plant Nutrition* (U.S.A.), are examples of such journals. A full list can be obtained from the Commonwealth Bureau of Soil Science, which has its headquarters at Rothamsted Experimental Station, Harpenden, England. The Commonwealth Bureau of Soil Science is itself responsible for the publication of *Technical Communications*, which are monographs on special topics as well as providing a reference service which will be referred to in more detail below.

The Society of Chemical Industry (14 Belgrave Square, London S.W.1.) publish a journal, *The Science of Food and Agriculture*, which contains soil papers and there are many published reports of symposia and conferences which are privately sponsored, and these should be found in the larger libraries or can be obtained for the benefit of readers by most librarians through the inter-library loan scheme.

Abstracts, Résumés, etc.

The Commonwealth Bureau of Soil Science is a division of a larger organisation, the Commonwealth Agricultural Bureau, sponsored by the member countries of the British Commonwealth to assist the work of agricultural scientists. This it does by reviewing all relevant scientific publications that are publicly available and publishing epitomes for the benefit of its subscribers. The monthly journal *Soils and Fertilisers* serves the soil worker and in it there are to be found, collected under subject headings, abstracts of papers designed to give an indication of the scope of the work. Each issue also contains one or more review articles, each on a single subject. In addition to its monthly journal the Bureau of Soil Science publishes *Technical Communications* as previously noted, and at three- or four-year intervals it issues a *Bibliography of Soil Science, Fertilisers and General Agronomy*. These books contain the titles and authors of all papers reviewed by the bureau during the period covered by the volume, the papers being fully indexed as to subject matter as inferred from their titles.

These general publications of the Commonwealth Bureau of Soil Science are, in some cases, adequate guides to the literature, but there are also other useful sources of information concerning published work. The Society of Chemical Industry publishes a valuable series of abstracts. These, formerly published in association with the Chemical Society of London, now appear as a section of the *Journal of the Science of Food and Agriculture*.

The American Chemical Society publishes *Chemical Abstracts*, and section 15 of this is devoted to papers relating to soils and fertilisers.

In addition to these detailed reports of published works there are several periodicals which offer more generalised surveys of the subject, some of which are well documented. The *Annual Reports of Applied Chemistry* has a section devoted to agriculture; the monthly journal of the Ministry of Agriculture, Fisheries and Food, *Agriculture*, often contains articles of interest; the *Proceedings of the Royal Agricultural Society* (annually) contains a report on recent work on soil, and many others also exist which may be encountered whilst browsing through a well-stocked library.

The Use of Literature

It has been indicated at the beginning of this chapter that there are two general reasons for the serious student of scientific agriculture necessarily having a general familiarity with the literature of the science of the soil. One is that he may have occasion to consider in some detail the work done on a particular problem, the other is that he must be able to keep in touch with the whole development of the subject. No 'rules' can be attempted for the achievement of these two ends, but the following general remarks may be useful.

Surveying the Literature of a Specific Problem

The student who wishes to ascertain whether any investigations have been carried out on a particular problem, and if so, with what results, will, in the first instance, naturally turn to such books on the soil as are available, and will search the references in *Soil Conditions and Plant Growth*. The *Bibliography of Soil Science, Fertilisers and Agronomy* will be consulted to find titles of papers which may have a bearing on the problem. This is not always a good and complete guide, as the subject of soil science is so diverse that not all the references may be grouped together in the manner desired, and in using the bibliography every possible heading under which references to the particular problem might be found has to be considered to find titles of papers and the reference to abstracts in *Soils and Fertilisers*. Many paper titles, too, which look promising often turn out to be quite irrelevant to the question under review and much time is often lost following up false clues. The abstracting journals are to be consulted in the initial stages of search, starting with the most recent issues and working backwards as far as is thought necessary. In using the index of an abstract journal care has to be taken to think of every possible word under which the subject might be indexed, for it is easily possible to miss the reference one wants because the indexer has put it under a word which does not occur to the reader. So far as is possible the original papers abstracted should be seen, as abstracts frequently give a wrong impression of the scope of the work. It is not *always* necessary to read them in every detail, as many papers have a few pages of summary and conclusions at the end.

When searching for papers dealing with a particular problem, there often occurs a paper which gives a résumé of work done on that problem up to the date of its acceptance for publication, and the student can often make use of the work of others.

The particular procedure adopted, however, depends upon the actual problem and the extent to which acquaintance with its detail is required. Experience of using the literature is the most effective way of indicating how to proceed in respect of a particular problem, and these remarks are made in the hope of stimulating the student to acquire some such experience.

Surveying the General Development of Soil Science

For this purpose it is still less possible to indicate a systematic procedure. In the agricultural periodicals mentioned above there frequently appear résumés of recent developments and at least some of these journals should be perused regularly. The *Annual Reports of Applied Chemistry* are useful in this connection and the monthly articles in *Soils and Fertilisers* give a cover of the problems and developments of more current interest. The non-specialist who wishes to be generally well informed should take such opportunities as present themselves to look through the later issues and editions of some of the journals and books with which he became acquainted during the period of systematic study.

Learned Societies

Whilst it is not perhaps strictly relevant in a chapter devoted to the literature of soil science to refer to the several learned societies related to the soil, an excuse can be made in that these organisations do, more often than not, publish much of value to the student and research worker.

There exist a number of societies with the aims and objects of furthering the frontiers of knowledge by the dissemination of information through paper readings and discussions. Membership of most such societies is open to all who have a genuine interest in soil science, and membership fees are usually modest. These societies operate through meetings of members and via their publications.

The British Society of Soil Science is of most direct concern in this country, being responsible for the assemblage and editing of the papers published in *The Journal of Soil Science*. In addition to this, it holds meetings for the discussion of soil researches in spring, and in the autumn it organises a field meeting with which is combined paper readings and discussions.

The Agriculture Group of the Society of Chemical Industry is responsible for the organisation of meetings held throughout the year, at many of which some aspect of soil science is the principal theme. This group is responsible for the preparation of such soil reviews as appear in the *Annual Reports of Applied Chemistry*.

Although primarily organised for the benefit of those engaged in educational work in its widest sense, the Agricultural Education Association is a society of long standing which meets twice yearly, in London during the winter and at some provincial centre during the summer. Among its deliberations, problems of the soil frequently occur and are particularly and generally discussed. It publishes the journal *Agricultural Progress*.

The International Society of Soil Science in its present form dates back to 1947, and membership is available to all interested at a nominal fee. Its roots go back to 1911 when the 'Internationale Mitteilungen' was initiated mainly by European soil chemists. In 1924 it was reconstituted as the International Society of Soil Science and its journal became the *Proceedings of the International Society of Soil Science*. The association disbanded with the outbreak of World War II and the journal was discontinued. The present society organises international conferences and restricts its publications to the proceedings of such conferences and to new bulletins.

Index